INTEGRATED PHOTONICS

INTEGRATED PHOTONICS: FUNDAMENTALS

Ginés Lifante
Universidad Autónoma de Madrid, Spain

WILEY

Telephone (+44) 1243 779777

Email (for orders and customer service enquiries): cs-books@wiley.co.uk
Visit our Home Page on www.wileyeurope.com or www.wiley.com

Other Wiley Editorial Offices

John Wiley & Sons Inc., 111 River Street, Hoboken, NJ 07030, USA

Jossey-Bass, 989 Market Street, San Francisco, CA 94103-1741, USA

Wiley-VCH Verlag GmbH, Boschstr. 12, D-69469 Weinheim, Germany

John Wiley & Sons Australia Ltd, 33 Park Road, Milton, Queensland 4064, Australia

John Wiley & Sons (Asia) Pte Ltd, 2 Clementi Loop #02-01, Jin Xing Distripark, Singapore 129809

John Wiley & Sons Canada Ltd, 22 Worcester Road, Etobicoke, Ontario, Canada M9W 1L1

Wiley also publishes its books in a variety of electronic formats. Some content that appears in print may not be available in electronic books.

Library of Congress Cataloging-in-Publication Data

Lifante, Ginés.
Integrated photonics : fundamentals / Ginés Lifante.
p. cm.
ISBN 0-470-84868-5
1. Photonics. I. Title.

TA1520 .L54 2002
621.36—dc21

2002191051

British Library Cataloguing in Publication Data

A catalogue record for this book is available from the British Library

ISBN 0-470-84868-5

Typeset in 10/12pt Times by Laserwords Private Limited, Chennai, India

To Victoria and Pepe

CONTENTS

PREFACE

If the last century was the era of electronics, the twenty-first century is probably the era of photonics. In particular, the miniaturisation of optical components will play an important role in the success of advanced photonic devices, based on optical waveguides. This book presents the basic concepts of waveguides necessary to understand and describe integrated photonic devices, from Maxwell's equations to the modelling of light propagation in arbitrary guiding structures.

The topics, as well as their depth of analysis in the book, have been established, benefiting from the experience of several years teaching this subject at the Universidad Autónoma de Madrid. Since integrated photonic devices have applications in very different areas, such as optical communication, environmental monitoring, biological and chemical sensing, etc., students following this course may have different backgrounds. Therefore, after the introductory chapter devoted to presenting the main characteristics of integrated photonic technology, in Chapter 2 we review the electromagnetic theory of light. In it the basis of electromagnetic waves is described, emphasising the most relevant concepts connected to optical waveguides, such as the phenomenon of total internal reflection.

Subsequent chapters deal with the fundamentals of integrated photonics: the theory of optical waveguides, the coupling mode theory and light propagation in guiding structures. Although the treatment given to the different topics is based upon fundamental principles, numerical examples based on real situations are given throughout, which permit the students to relate theory to practice.

I am indebted to Professor F. Cussó, who encouraged me to write this book. I would like also to thank Professor I. Aguirre and Professor J.A. Gonzalo who carefully read the manuscript, and to Professor F. Jaque, in particular, who helped me with his invaluable suggestions.

I also want to express my very special appreciation to A. Bagney for her kind help in correcting and preparing the book in its final form.

ABOUT THE AUTHOR

Ginés Lifante Pedrola, a native of Jumilla (Spain), is a graduate of the Universidad Autónoma de Madrid. After a master's degree completed with a thesis on "Luminescent Solar Concentrators", he received his PhD under the direction of Professor F. Cussó, with a thesis on the topic of "Materials for Colour Centre Lasers". He has undertaken research study at Strathclyde University, Glasgow, with Professor B. Henderson working on colour centres lasers, at Sussex University, working with Professor P.D. Townsend doing theoretical and experimental research on non-linear waveguides made by ion implantation, and at CNRS-LAAS, Toulouse, working with Dr A. Muñoz-Yagüe on active waveguides grown by MBE using UV transparent materials.

His present research topic is the field of integrated photonic devices based on active and functional materials with applications in optical communication technology and environmental sensing. He is in charge of several projects in this field, is co-author of a hundred papers, and has several patents.

Professor Lifante has a broad teaching experience covering different teaching levels, including optics, optoelectronics and integrated photonics, and has directed several doctoral theses on integrated optics.

When not working, he is the respected coach of the Soccer Physics Team at the UAM.

1

INTRODUCTION TO INTEGRATED PHOTONICS

Introduction

The term "integrated photonics" refers to the fabrication and integration of several photonic components on a common planar substrate. These components include beam splitters, gratings, couplers, polarisers, interferometers, sources and detectors, among others. In turn, these can then be used as building blocks to fabricate more complex planar devices which can perform a wide range of functions with applications in optical communication systems, CATV, instrumentation and sensors. The setting-up of integrated photonic technology can be considered as the confluence of several photonic disciplines (dealing with the control of light by electrons and vice versa) with waveguide technology. In fact, optical waveguides are the key element of integrated photonic devices that perform not only guiding, but also coupling, switching, splitting, multiplexing and demultiplexing of optical signals. In this chapter we will introduce the main characteristics of integrated photonic technology, showing relevant aspects concerning material and fabrication technologies. Also, we will briefly describe some basic components present in integrated photonic devices, emphasising the differences in their design compared to conventional optics. Some examples of integrated photonic devices (passive, functional, active and non-linear) are given at the end of the chapter to show the elegant solution that this technology proposes for the development of advanced optical devices.

1.1 Integrated Photonics

Optics can be defined as the branch of physical science which deals with the generation and propagation of light and its interaction with matter. Light, the main subject of optics, is electromagnetic (EM) radiation in the wavelength range extending from the vacuum ultraviolet (UV) at about 50 nanometers to the far infrared (IR) at 1 mm. In spite of being a very ancient science, already studied by the founder of the School of Alexandria, Euclid, in his *Optics* (280 BC), during the last quarter of the past century, the science of optics has suffered a spectacular renaissance, due to various key developments. The first revolutionary event in modern optics was, no doubt, the invention of the laser by T.H. Maiman in 1960 at Hughes Research Laboratories in Malibu [1], which allowed the availability of coherent light sources with exceptional properties, such as high spatial and temporal coherence and very high brightness. A second major step forward came with the development of semiconductor optical devices for the generation and detection of light, which permitted very efficient and compact opto-electronic devices. The last push was given by the introduction of new fabrication techniques for obtaining very cheap optical fibres, with very low propagation losses, close to the theoretical limits (Figure 1.1).

As a result of these new developments and associated with other technologies, such as electronics, new disciplines have appeared connected with optics: electro-optics, opto-electronics, quantum electronics, waveguide technology, etc. Thus, classical optics, initially dealing with lenses, mirrors, filters, etc., has been forced to describe a new family of much more complex devices such as lasers, semiconductor detectors, light modulators, etc. The operation of these devices must be described in terms of optics as well as of electronics, giving birth to a mixed discipline called *photonics*. This new discipline emphasises the increasing role that electronics play in optical devices, and also the necessity of treating light in terms of photons rather than waves, in particular in terms of matter–light interactions (optical amplifiers, lasers, semiconductor devices, etc.). If electronics can be considered as the discipline that describes the flow of electrons, the term "photonics" deals with the control of photons. Nevertheless, these two disciplines clearly overlap in many cases, because photons can control the flux of electrons, in the case of detectors, for example, and electrons themselves can determine the properties of light propagation, as in the case of semiconductor lasers or electro-optic modulators.

The emergence of novel photonic devices, as well as resulting in the important connection between optics and electronics, has given rise to other sub-disciplines within photonics. These new areas include electro-optics, opto-electronics, quantum optics, quantum electronics and non-linear optics, among others. *Electro-optics* deals with the study of optical devices in which the electrical interaction plays a relevant role in controlling the flow of light, such as electro-optic modulators, or certain types of lasers. *Acousto-optics* is the science and technology concerned with optical devices controlled by acoustic waves, driven by piezo-electric transducers. Systems which involve light

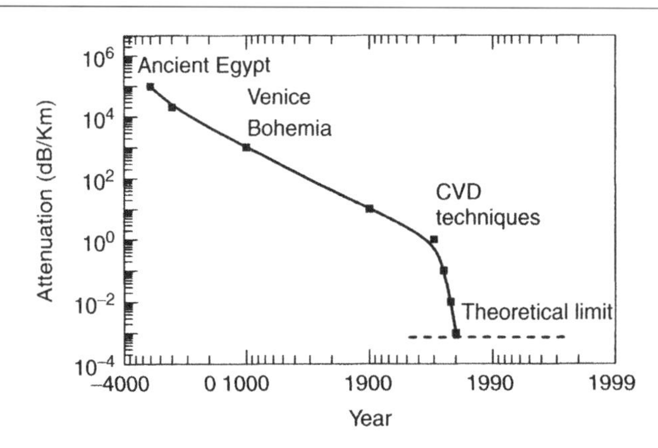

Figure 1.1 Evolution of the attenuation in silica glasses. In the 1980s the dramatic drop in the attenuation coincides with the boom of the optical fibre systems, thanks to the implementation of new fabrication techniques

but are mainly electronic fall under *opto-electronics*; these systems are in most cases semiconductor devices, such as light-emitting diodes (LEDs), semiconductor lasers and semiconductor-based detectors (photodiodes). The term *quantum electronics* is used in connection with devices and systems that are based on the interaction of light and matter, such as optical amplifiers and wave-mixing. The quantum nature of light and its coherence properties are studied in *quantum optics*, and the processes that involve non-linear responses of the optical media are covered by the discipline called *non-linear optics*. Finally, some applied disciplines emerging from these areas include *optical communications*, *image* and *display systems*, *optical computing*, *optical sensing*, etc. In particular, the term *waveguide technology* is used to describe devices and systems widely used in optical communications as well as in optical computing, optical processing and optical sensors.

A clear example of an emergent branch of optics that combines some of the above disciplines is the field of *integrated optics*, or more precisely, *integrated photonics*. We consider integrated photonics to be constituted by the combining of waveguide technology (guided optics) with other disciplines, such as electro-optics, acousto-optics, non-linear optics and opto-electronics (Figure 1.2). The basic idea behind integrated photonics is the use of photons instead of electrons, creating integrated optical circuits similar to those in conventional electronics. The term "integrated optics", first proposed in 1960 by S.E. Miller [2], was introduced to emphasise the similarity between planar optical circuits technology and the well-established integrated micro-electronic circuits. The solution proposed by Miller was to fabricate integrated optical circuits through a process in which various elements, passive as well as active, were integrated in a single substrate, combining and interconnecting them via small optical transmission lines called waveguides. Clearly, integrating multiple optical functions in a single photonic device is a key step towards lowering the costs of advanced optical systems, including optical communication networks.

The optical elements present in integrated photonic devices should include basic components for the generation, focusing, splitting, junction, coupling, isolation, polarisation control, switching, modulation, filtering and light detection, ideally all of them

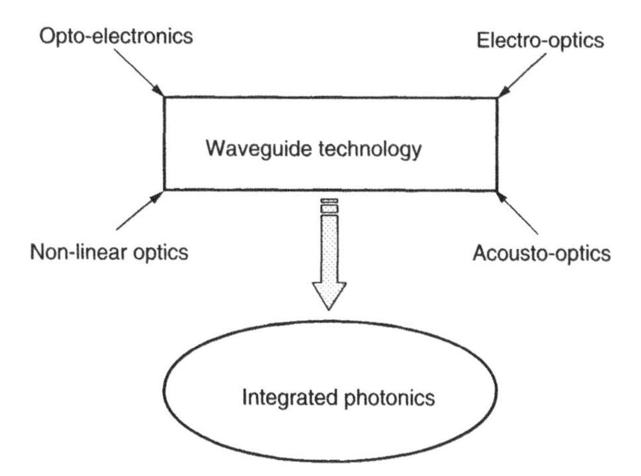

Figure 1.2 Confluence of various disciplines into integrated photonics

being integrated in a single *optical chip*. Channel waveguides are used for the interconnection of the various optical elements. The main goal pursued by integrated photonics is therefore the miniaturisation of optical systems, similar to the way in which integrated electronic circuits have miniaturised electronic devices, and this is possible thanks to the small wavelength of the light, which permits the fabrication of circuits and compact photonic devices with sizes of the order of microns. The integration of multiple functions within a planar optical structure can be achieved by means of planar lithographic production [3]. Although lithographic fabrication of photonic devices requires materials different from those used in microelectronics, the processes are basically the same, and the techniques well established from 40 years of semiconductor production are fully applicable. Indeed, a lithographic system for fabricating photonic components uses virtually the same set of tools as in electronics: exposure tools, masks, photoresists, and all the pattern transfer process from mask to resist and then to device.

1.2 Brief History of Integrated Photonics

For 30 years after the invention of the transistor, the processing and transmission of information were based on electronics that used semiconductor devices for controlling the electron flux. But at the beginning of the 1980s, electronics was slowly supplemented by and even replaced by optics, and photons substituted for electrons as information carriers. Nowadays, photonic and opto-electronic devices based on integrated photonic circuits have grown in such a way that they not only clearly dominate long-distance communications through optical fibres, but have also opened up new fields of application, such as sensor devices, and are also beginning to penetrate in the own field of the information processing technology. In fact, the actual opto-electronic devices may be merely a transition to a future of all-optical computation and communication systems.

The history of integrated photonics is analogous to that of other related technologies: discovery, fast evolution of the devices, and a long waiting time for applications [4]. The first optical waveguides, fabricated at the end of the 1960s, were bidimensional devices on planar substrates. In the mid-1970s the successful operation of tridimensional waveguides was demonstrated in a wide variety of materials, from glasses to crystals and semiconductors. For the fabrication of functional devices in waveguide geometries, lithium niobate ($LiNbO_3$) was rapidly recognised as one of the most promising alternatives. The waveguide fabrication in $LiNbO_3$ via titanium in-diffusion was demonstrated at the AT&T Bell Laboratory, and gave rise to the development of channel waveguides with very low losses in a material that possesses valuable electro-optic and acousto-optic effects. In the mid-1980s the viability of waveguide devices based on $LiNbO_3$, such as integrated intensity modulators of up to 40 GHz, and with integration levels of up to 50 switches in a single photonic chip had already been demonstrated in laboratory experiments. A few years later, the standard packaging required in telecommunication systems was obtained, and so the devices were ready to enter the market. The rapid boom of monomode optical fibre systems which started in the 1980s was the perfect niche market for these advanced integrated photonic devices that were waiting in the research laboratories. Indeed, the demand for increased transmission capacity (bandwidth) calls urgently for new integrated photonic chips that permit the control and processing of such huge data transfer, in particular with the introduction of technology to transmit light in multiple wavelengths (WDM, wavelength division multiplexing).

Because of the parallel development of other materials, both dielectrics such as polymers, glasses or silica on silicon (SiO_2/Si), and semiconductors such as indium phosphide (InP), gallium arsenide (GaAs) or even silicon (Si), a wide variety of novel and advanced integrated photonic devices was ready to emerge on the market. During the last two decades of the twentieth century we have moved from the development of the new concept of integrated optical devices to a huge demand for such novel devices to implement sophisticated functions, mainly in the optical communication technology market. In fact, at the beginning of the twenty-first century the data transfer created by computer-based business processes and by Internet applications is growing exponentially, which translates into a demand for increasing transmission capacity at lower cost, which can only be met by increased use of optical fibre and associated advanced photonic technologies (Figure 1.3). Today fibres are typically used to transmit bit-rates up to 10 Gbit/s, which is, however, far below the intrinsic bandwidth of an optical fibre. Wavelength Division Multiplexing (WDM) (the transmission of several signals through a single fibre using several wavelengths) paves the way to transmit information over an optical fibre in a much more efficient way, by combining several 10 Gbit/s signals on a single fibre. Today there are commercial WDM systems available with bit-rates in the range of 40 to 400 Gbit/s, obtained by combining a large number of 2.5 and 10 Gbit/s signal, and using up to 32 different wavelengths. The next frontier in data transfer capacity points to the Terabit transmission, which can be achieved by using Time Domain Multiplexing (TDM), an obvious multiplexing technique for digital signals. An equivalent of TDM in the optical domain (OTDM) is also being developed with the purpose of reaching much higher bit-rates which will require the generation and transmission of very short pulses, in the order of picoseconds, and digital processing in the optical domain. Clearly, all these technologies will require highly advanced optical components, and integrated photonic devices based on planar lightwave circuits are the right choice to meet the high performance levels required, which allow the integration of multiple functions in a single substrate (Table 1.1).

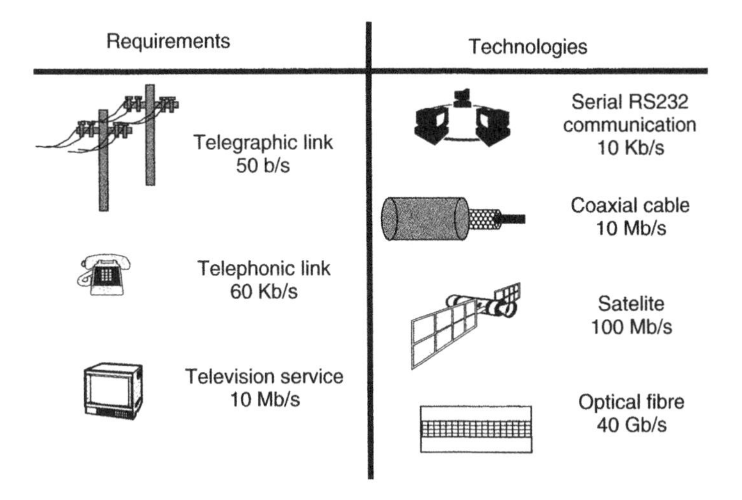

Figure 1.3 Requirements for data transfer and available technologies

Table 1.1 Integrated optics market in 2001 by material type [5]

Material	%
Lithium niobate	30
Indium phosphide	22
Gallium arsenide	20
Silica on silicon	11
Polymer	5
Silicon	3
Other	9
TOTAL	100

1.3 Characteristics of the Integrated Photonic Components

The basic idea behind the use of photons rather than electrons to create integrated photonic circuits is the high frequency of light (200 THz), which allows a very large bandwidth for transporting and managing a huge amount of information. The replacement of electronic by photonic means is forced by fundamental physical reasons that limit the information transmission rate using purely electronic means: as the frequency of an electrical signal propagating through a conductor increases, the impedance of the conductor also increases, thus the propagation characteristics of the electrical cable become less favourable. That is the reason why electrical signals with frequencies above 10 MHz must be carried by specially designed conductors, called coaxial cables, in order to minimise the effect of a high attenuation. Figure 1.4 shows the attenuation in a typical coaxial cable as a function of the frequency. It can be seen that for high transmission rates (~100 MHz), the attenuation is so high (~5 dB/Km) that communications based on electrical signals propagating on coaxial cables can be used in applications where the typical distances are tens of metres (buildings), but they are

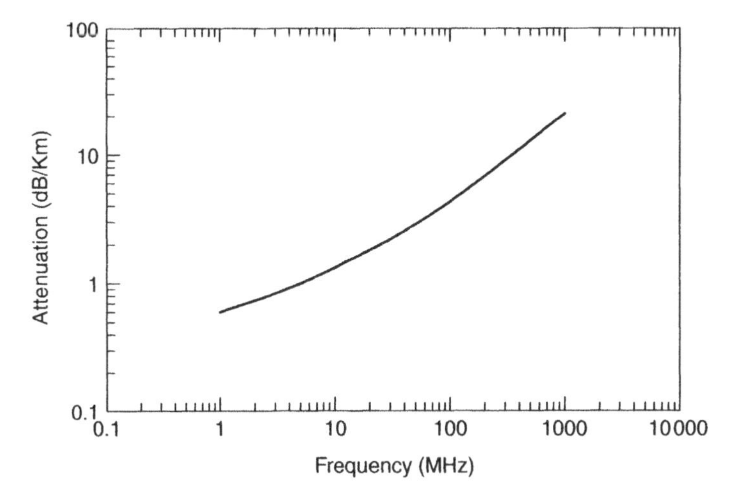

Figure 1.4 Attenuation of the electrical signal in a coaxial cable as a function of the modulation frequency

useless for distances greater than several kilometres (links between cities). In contrast, optical signals propagate through non-conducting dielectric media, operating in the wavelength range where the materials are highly transparent. For most optical materials used in optical communications and photonic devices, this transparent window falls in the visible and near-infrared range of the electromagnetic spectrum, which corresponds to light frequency in the range 150–800 THz, 10^6 times the frequency used in electrical transmission!

Integrated photonic devices based on integrated optical circuits take advantage of the relatively short wavelength of the light in this range (0.5–2 μm), which allows the fabrication of miniature components using channel waveguides the size of microns. The technology required to fabricate planar lightwave circuit components of such dimensions is therefore common in the well-established Micro-electronic technology, using the tools and techniques of the semiconductor industry.

The basic concept in optical integrated circuits is the same as that which operates in optical fibres: the confinement of light. A medium that possesses a certain refractive index, surrounded by media with lower refractive indices, can act as a light trap, where the rays cannot escape from the structure due to the phenomena of total internal reflection at the interfaces. This effect confines light within high refractive index media, and can be used to fabricate optical waveguides that transport light from point to point, whether long distances (optical fibres) or in optical circuits (integrated photonic chips). Figure 1.5 shows the basic structures for the most common waveguide geometries. In a planar waveguide (Figure 1.5a) light is trapped by total internal reflection in a film (dashed region), and therefore the film must have a refractive index greater than the refractive indices corresponding to the upper and lower media. These are usually referred to as the *cover* and the *substrate*, respectively, and the film is called the *core*, because that is where most of the optical energy is concentrated.

In a channel waveguide the light propagates within a rectangular channel (the dashed region in Figure 1.5b) which is embedded in a planar substrate. To confine light within the channel it is necessary for the channel to have a refractive index greater than that of the substrate, and of course, greater than the refractive index of the upper medium, which is usually air. This type of waveguide is the best choice for fabricating integrated photonic devices. Because the substrate is planar, the technology associated with integrated optical circuits is also called *planar lightwave circuits* (PLC).

Finally, Figure 1.5c shows the geometry of an optical fibre, which can be considered as a cylindrical channel waveguide. The central region of the optical fibre or core is surrounded by a material called *cladding*. Of course, the core must have a higher

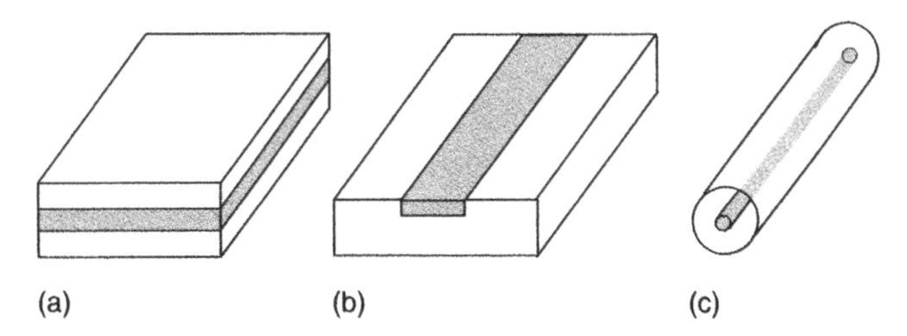

Figure 1.5 Basic waveguide geometries: (a) planar waveguide; (b) channel waveguide; (c) optical fibre

refractive index than the cladding in order to trap light within the structure after total internal reflection.

In both channel waveguides and optical fibres the confinement of optical radiation takes place in two dimensions, in contrast to planar waveguides where there is only light confinement in a single direction. This fact allows light in planar waveguides to diffract in the plane of the film, whereas in the case of channel waveguides and fibres diffraction is avoided, forcing the light propagation to occur only along the structure's main axis.

Three generations can be distinguished in the evolution of optical systems, from conventional optical systems to integrated optical circuits (Table 1.2). The first generation concerns *conventional optical* systems, where the optical components with sizes of the order of centimetres were set on optical benches typically with dimensions of metres, while the optical beams had diameters of the order of several millimetres. A second generation in the evolution of optical systems can be called *micro-optics*. Its main characteristic is the use of miniature optical components such as light emitting diodes, diode lasers, multi-mode fibres, etc. These components are clearly a transition towards the devices used nowadays in modern communication systems based on optical fibres. Nevertheless, although the characteristics of micro-optic systems are satisfactory, there are problems with the alignment and coupling between the components because of their small size (of the order of millimetres). Furthermore, because of the critical alignment, the various optical components are not packed together, making the optical system unstable. The last generation in optical systems concerns integrated photonics, and is based on optical circuits and components integrated in a single substrate. This, as well as the small size of the optical components, is the key factor for the success of integrated photonic systems. This technology, with unique features with respect to previous generations, possesses important advantages in terms of choice of materials, design, fabrication and performance characteristics. Some of the special features of systems based on integrated photonic technology are the following:

1. *Functionality based on electromagnetic optics*. The key elements in an integrated optical device are monomode channel waveguides with width and depth typically of the order of microns, where the optical radiation propagates in a single mode. In this way, while the optical systems of the first and second generation can be adequately

Table 1.2 Evolution of the optical systems technology and relevant characteristics [6]

	First generation	Second generation	Third generation
Technology	Conventional optics	Micro-optics	Integrated photonics
Typical components	Mirrors, prisms, lenses, gas lasers	LED, LD, tiny lenses, multi-mode fibres	Monomode channel waveguides, LD, monomode fibres
Alignment	Necessary	Necessary (hard)	Unnecessary
Propagation	Beam	Beams in multi-mode waveguides	Monomode waveguides (μm)
Control electrode size	1 cm	1 mm	1 μm
Device size	1 m^2	10 cm^2	1 cm^2 (on a 1 mm thick substrate)

Note: LED: light emitting diode; LD: laser diode.

treated by ray optics because of the wide diameters of the optical beams (compared to the wavelength of the light), integrated optical devices must be analysed considering the propagating light as electromagnetic waves.

2. *Stable alignment.* A key factor in the good performance of an optical system is the adjustment and alignment of the various optical components, which is critical and difficult to achieve for conventional optical systems. In contrast, in integrated photonic devices, once the optical chip has been fabricated, the alignment problem is avoided and stability is assured. Furthermore, the device is stable against vibrations or thermal changes. This characteristic, which is the most relevant feature in integrated photonic devices, is assured because all the optical elements are integrated in a single substrate.
3. *Easy control of the guided modes.* Because the waveguides are monomode, it is easier to control the optical radiation flux through the electro-optic, acousto-optic, thermo-optic or magneto-optic effects, or even by the light itself via non-linear interactions. If the waveguides were multi-mode, this control by external fields would be much more complicated, because of the different propagation characteristics of each modal field.
4. *Low voltage control.* For devices based on light control via the electro-optic effect, the short width of the channel waveguides allows one to drastically reduce the distance between the control electrodes. This implies that the voltage required to obtain a certain electric field amplitude can be considerably reduced. For example, while the typical voltage for electro-optic control in conventional optical systems is of the order of several KV, in integrated optic devices the voltage required is only a few volts.
5. *Faster operation.* The small size of the control electrodes in an electro-optic integrated photonic device implies low capacitance, and this allows for a faster switching speed and higher modulation bandwidth. Typical modulations of 40 Gbit/s are easily achieved using lithium niobate, polymers or InP-based devices.
6. *Effective acousto-optic interactions.* Since the field distributions of surface acoustic waves (SAW) are located within a distance of a few wavelengths beneath the substrate surface (tens of microns), the SAW and the optical waveguide modes overlap strongly, giving rise to efficient acoustooptic interactions. Thus, using SAW generated by piezotransducers, high performance integrated optical devices based on acoustooptic effect can be developed.
7. *High optical power density.* Compared with conventional optical beams, the optical power density in a monomode channel waveguide is very high, due to the small cross-sectional area of the guide. This is of special relevance in the performance of devices requiring high radiation intensity, such as frequency converters (via non-linear effects) or even optical amplifiers and lasers. These devices are therefore very efficient when designed and fabricated with integrated photonic technology.
8. *Compact and low weight.* The use of a single substrate with an area of several millimetres squared for integrating different photonic components makes the optical chip very compact and very light weight.
9. *Low cost.* The development of integration techniques makes mass production possible via lithographic techniques and mask replication; also, the planar technology reduces the quantity of material necessary to fabricate the photonic devices. These

aspects are the basis of a low cost device and thus an easy introduction into the market.

1.4 Integrated Photonics Technology

The technology and fabrication methods associated with integrated optical circuits and components are very varied, in addition, they depend on the substrate material with which the optical device is fabricated. The methods most widely used in the definition of optical circuits over a substrate are diffusion techniques (such as titanium diffusion in lithium niobate) and deposition techniques (such as chemical vapour deposition used for silica). Since the lateral dimensions of the optical circuits are only a few microns, the fabrication technology needs photolithographic processes. In the case of diffusion techniques it is possible to use photolithographic masks to define open channels through which the diffused material enters the substrate, or, alternatively, one can deposit the previously patterned material to be diffused directly onto the substrate. For waveguides fabricated by deposition techniques the lateral definition of the optical circuits is usually carried out by means of etching after the deposition of the material onto the whole substrate surface.

Optical integration can expand in two directions: serial integration and parallel integration. In serial integration for optical communication devices the different elements of the optical chip are consecutively interconnected: laser and driver, modulator and driver electronics, and detector and receiver electronics. In parallel integration, the chip is built by bars of amplifiers, bars of detectors and wavelength (de)multiplexors. Also, a combination of these two architectures should incorporate optical cross-connects and add-drop modules. The highest level of integration (whether serial or parallel) is achieved in monolithic integration, where all the optical elements including light sources, light control, electronics and detectors are incorporated in a single substrate. The most promising materials to achieve full monolithic integration are semiconductor materials, in particular GaAs and InP. In hybrid integration technology, the optical chip fabricated on a single substrate controls the optical signals, while additional elements such as lasers or detectors are built on different substrates and are directly attached to the integrated photonic device or interconnected by optical fibres. Examples of hybrid technologies include dielectric substrates, such as glasses, silica or ferro-electric crystals. The case of silica on silicon can be considered as quasi-hybrid integration, in the sense that optical components, electronics and detector can be implemented in a single substrate, but not the light source.

All integrated photonic devices require input/output optical signals carried by optical fibres. Indeed, one of the most difficult tasks in packaging an integrated optical device is attaching the fibres to the chip waveguides, known as fibre pigtails. The fibre alignment is typically 0.1 micron or less for low power loss, where the optical chip surfaces should be carefully polished at odd angles to eliminate back reflection from the interface. This alignment must be maintained during the attachment and also through subsequent thermal transitions as well as in shock and vibration-prone environments while the device is operating.

Lithography replicates a prototype from chip to chip or from substrate to substrate. Although a lithographic system for fabricating photonic devices uses the same tools as in semiconductor electronics, there are some important differences. First, while in electronics, bends and interconnections affect the maximum data rates, in photonic circuits the major impact is on optical power throughput. Second, while electrons strongly interact with each other, photons can exist even in the same circuit without interacting. As a consequence, integrated circuits in electronics usually have an overall square geometry, with multiple layers to enable the cross-over of electrical signals, while integrated optical chips tend to have a single layer and an elongated geometry with unidirectional flow to minimise bending of the optical path.

Although there is a great number of lithographically processable materials that can be used to fabricate optical waveguides, only a few of them have shown the required characteristics to develop integrated optical devices. These include a wide range of glasses, crystals and semiconductors (Table 1.3). In particular, the substrates most commonly used are glasses, lithium niobate, silica on silicon, III-V semiconductor compounds and polymers. Each type of material has its own advantages and disadvantages, and the choice of a specific substrate depends on the particular application of the photonic device. Nowadays there exists a great variety of devices based on each of these materials.

The glass-based integrated optical devices have the great advantage of the low cost of the starting material and the fabrication technique, mainly performed by an ionic exchange process [7]. The method used for producing waveguides in glass substrates

Table 1.3 Materials technology for integrated photonic devices

Substrate	Material properties	Waveguide technology	Advantages	Demonstrated devices
Multi-components Glasses	Low price Rare earths incorporation	Ionic exchange	Easy and cheap fabrication Low losses	Passive devices Amplifiers
SiO_xN_y:SiO_2:Si $TiO_2/SiO_2/Si$	Cheap and versatile fabrication	Thermal oxidation CVD, FHD, ECR, Sol-gel	Versatility Microelectronic technology	Passive devices TO switches AWG
Lithium niobate	Electro-optic Acousto-optic Non-linear Bi-refringent	Metallic diffusion Protonic exchange	Easy control of light Anisotropic	Switches Modulators Couplers WDM and DWDM
III-V compounds (InP, GaAs)	Electro-optic Light source Light detection Electronics	Epitaxy (MBE, LPE, CVD, MOCVD)	High level of integration	Modulators Amplifiers Lasers AWG
Polymers	Electro-optics Thermo-optics Non linear	Spin coating Dip Coating	High versatility Wide range of physical properties	Chemical and biological sensors TO switches EO Modulators

Notes: CVD: chemical vapour deposition; FHD: flame hydrolysis deposition; ECR: electron cyclotron resonance; MBE: molecular beam epitaxy; LPE: liquid phase epitaxy; MOCVD: metal-organic chemical vapour deposition; TO: thermo-optic; AWG: arrayed waveguide grating; WDM: wavelength division multiplexing; DWDM: dense WDM; EO: electro-optic.

is the exchange of alkali ions from the glass matrix (usually Na^+ ions) for monovalent cations such as K^+, Ag^+, Cs^+ or Tl^+, immersing the glass substrate in a molten salt that contains some of these ions at temperatures in the range 200–500°C, depending on the type of glass and the particular salt. For defining the optical circuits, a stopping mask is deposited onto the substrate, in such a way that the ionic exchange takes place only in the channels opened in the mask. This mask is removed after the exchange process. The refractive index increase due to the ionic exchange depends both on the glass composition and on the exchanged ions, and typically varies in the range 0.01 to 0.1. Since the glasses are amorphous materials, they do not present physical properties useful for the direct control of light, and therefore they are used mainly for the fabrication of passive devices.

One of the materials most widely used in the fabrication of integrated optical devices is lithium niobate ($LiNbO_3$) [8]. This is due to several characteristics of this crystalline material. In the first place, $LiNbO_3$ presents very interesting physical properties: in particular, it has valuable acousto-optic, electro-optic and piezo-electric effects. These properties allow the fabrication of functional devices such as phase modulators, switches, directional couplers, multiplexors, etc. Besides being a birefringent material, $LiNbO_3$ shows high non-linear optical coefficients, and these two properties permit very efficient frequency conversion, such as second harmonic generation and optical parametric oscillation. Furthermore, several techniques for waveguide fabrication in $LiNbO_3$ are now well established, including Ti or Zn metallic diffusion, protonic exchange, or even ion implantation. The resulting waveguides have very low losses, typically in the range of 0.01–0.2 dB/cm. Integrated optical circuits technology based on $LiNbO_3$ substrates is now very well established, and a great variety of devices based on this technology, mainly in the field of optical communications, are now commercially available.

The main advantage of silica over silicon-based photonic waveguides is the low price and the good optical quality of the silicon substrates, besides being a well-known material with a long tradition, and the experience developed from micro-electronic technology. The first step in waveguide fabrication using silicon substrates is the deposition of a silicon dioxide layer a few microns thick, which can also be obtained by direct oxidation of the silicon at high temperature. This layer has a double purpose: to provide a low index region for allowing light confinement, and also to move away the highly absorbing silicon substrate. For this reason this layer is called a *buffer* layer. The waveguide core is formed by further deposition of a high index oxynitride layer, usually via the chemical vapour deposition method (CVD) or the flame hydrolysis deposition (FHD) method [9]. The refractive index of the oxynitride core, SiO_xN_y, can be continuously varied in the range 1.45–2.1 by controlling the relative concentration of SiO_2 and Si_3N_4 compounds during the deposition. As the SiO_2 buffer layer has a refractive index of 1.45, a very high index contrast between the waveguide core and the surrounding media can be obtained. The most appealing feature of silicon as a substrate in integrated photonics is the possibility of integrating the detector and the associated electronic in a single platform substrate.

Perhaps, second to $LiNbO_3$ the III–V semiconductor compounds (mainly GaAs and InP) are the substrates with greatest impact on integrated optics technology, and are probably the materials with the most promising future in this field [10, 11]. The importance of the III–V compounds in integrated photonics derives from the fact that they offer the possibility of a high level of monolithic integration. Indeed, InP is a very versatile platform that promises large-scale integration of active components (lasers and detectors), passive components, and also electronics. The electronic technology of these semiconductor materials is now well established, and optical waveguide fabrication is quite straightforward by modifying the dopant concentration during the deposition process, Al in the case of GaAs, and Ga or As in the case of InP. The main problem concerning this technology has its roots in the relatively high losses of waveguides made of these materials (>1 dB/cm). Nevertheless, the fabrication technology in InP is rapidly improving, and several integrated photonic devices that show very high performance are now available in the market, such as semiconductor optical amplifiers, arrayed waveguide gratings or high speed modulators.

Among the materials suitable for integrated photonic technology, polymers occupy a special position, due to the fact that they exhibit some very useful physical properties, such as electro-optic, piezo-electric and non-linear effects, with values even higher that those of lithium niobate crystals [12]. Also, the thermo-optic coefficient for polymers is more than ten times higher than the corresponding coefficients for silica. The waveguide fabrication method for polymers starts from a solution of the polymeric material, followed by a deposition by spin coating or dip coating on a substrate. Due to their easy processing, the polymer layers allow for great flexibility when choosing a substrate; they are compatible with very different substrates such as glasses, silicon dioxide, or even silicon and indium phosphide. The choice of a particular polymeric material should take into consideration some important properties such as high transparency, easy processing, and high physical, chemical, mechanical and thermal stability. The main advantage of polymer-based integrated optical devices is their high potential for use in the field of chemical and biological sensors, because the organic groups in the polymeric compound can be designed and tailored to react against a specific medium. Also, due to the large electro-optic coefficient showed by some polymers, high speed and low voltage switches and modulators have been developed for the telecommunication market, offering high performance at low cost.

1.5 Basic Integrated Photonic Components

As in electronics, in integrated photonics there are some basic components common to most of the integrated optical devices. Although in essence all these components basically perform the same functions as their corresponding devices in conventional optics, the operating principles are usually quite different, and thus their design has very little to do with traditional optical components.

Although nowadays a long list of integrated photonic devices has been proposed, modelled and fabricated, and their number is quickly increasing, the basic components remain almost unchanged. Therefore it is possible to describe a short list of such components, basic blocks from which much more complex integrated optical devices can be built. We will now briefly outline some of the most common components, and we will show the dramatic change in design concept of integrated photonic devices compared to conventional optical components performing the same function. The main difference in design comes from the fact that while in conventional optics the operation principle is based on the behaviour of the light considered as plane waves or rays, in integrated optics the modelling and performance of the devices should be treated using the formalism of electromagnetic waves; this is because the size of the beams is of the order of the light's wavelength, typically in the range of microns. In fact, optical propagation in integrated photonic devices is conveyed through optical channels with dimensions of a few micrometres, both in depth and in width. Channel waveguides are defined in a single plane substrate, and other related elements (electrodes, piezoelements, heaters, etc.) are mounted on the same substrate, giving rise to a robust and compact photonic device. Unless otherwise stated, all the basic components that we will now describe will be based on monomode channel waveguides.

All the optical components in integrated photonics are constructed with three building blocks. They are the straight waveguide, the bend waveguide and the power splitter. Using these building blocks, several basic components have been developed to perform basic optical functions. In addition, a particular function can be executed using different elements, whose design may differ substantially. This versatility in optical element conception is one of the special features of integrated photonic technology. Now, we shall discuss several of these basic blocks and optical elements that perform some basic functions common in many integrated optical devices.

- *Interconnect*. This basic element serves to connect optically two points of a photonic chip (Figure 1.6a). The straight channel waveguide (Figure 1.6b), being the simplest structure for guiding light, interconnects different elements which are aligned on the optical chip. It can also act as a spatial filter, maintaining a Gaussian-like mode throughout the chip architecture. In order to interconnect different elements which are not aligned with the optical axis of the chip, a bend waveguide is needed, and therefore a bend waveguide is often called an offset waveguide (Figure 1.6c). These are also used to space channel waveguides at the chip endfaces, so that multiple fibres may be attached to it.
- *Power splitter* 1×2. A power splitter 1×2 is usually a symmetric element which equally divides power from a straight waveguide between two output waveguides (Figure 1.6d). The simplest version of a power splitter is the Y-branch (Figure 1.6e), which is easy to design and relatively insensitive to fabrication tolerances. Nevertheless, the curvature radii of the two branches, as well as the junction, must be carefully designed in order to avoid power losses. Also, if the two branches are separated by tilted straight waveguides, the tilt angle must be small, typically a few degrees. A different version of a power splitter is the multi-mode interference element (MMI, Figure 1.6f). This name comes from the multi-modal character of the wide waveguide region where the power split takes place. The advantage of this design is the short length of the MMI compared to that of the Y-branch. Although the dimensions of the MMI are not critical, allowing wide tolerances, this element must be designed for a particular wavelength. The two power splitters which have been described are symmetric, and thus 50% of the input power was carried by each output waveguide. Nevertheless, asymmetric splitters can also be designed for specific purposes. In addition, it is possible to fabricate splitters with N output waveguides, and in that case the element is called a $1 \times N$ splitter.
- *Waveguide reflector*. The waveguide reflector performs the task of reflecting back the light in a straight waveguide (Figure 1.6g). The simplest method of performing this task is to put a metallic mirror at the end of the channel waveguide (Figure 1.6h). If one needs the reflection to occurs only for a particular wavelength, a multistack dielectric mirror is used. Another way of building a waveguide reflector is

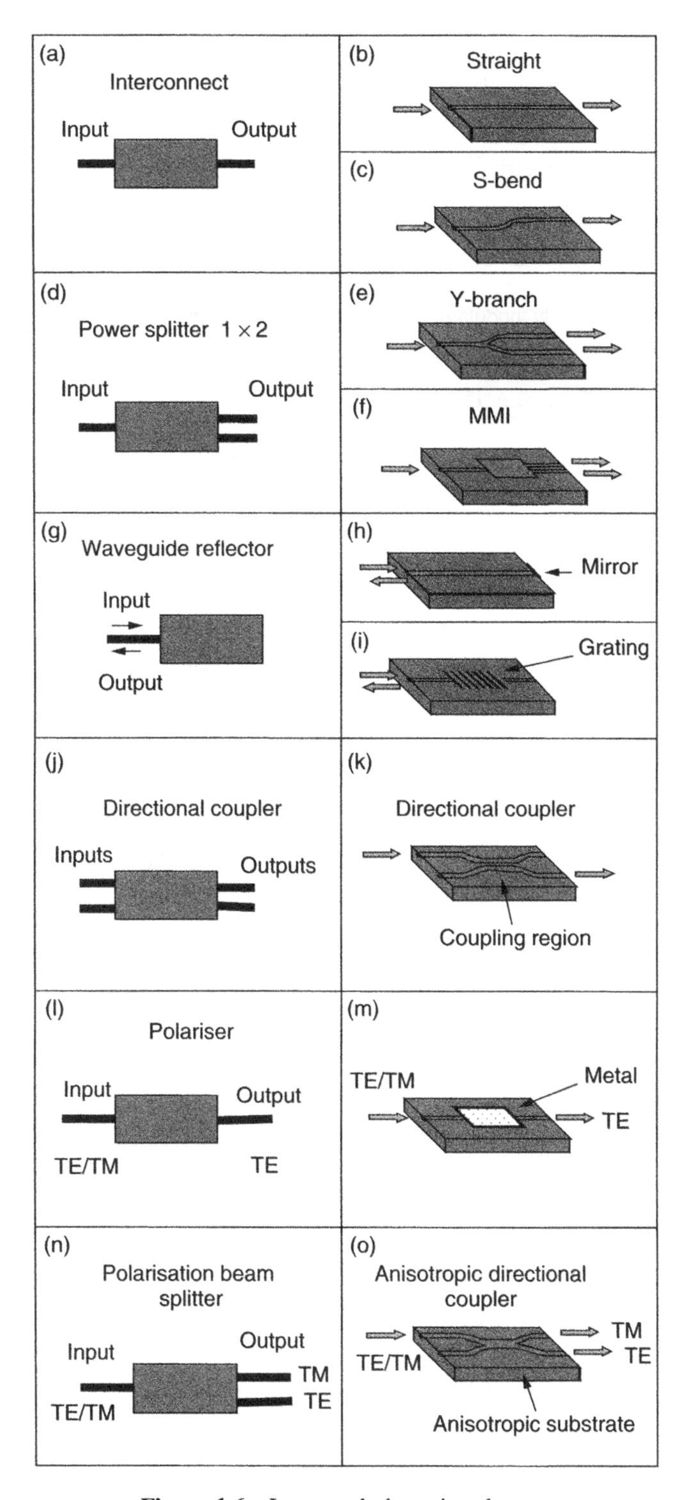

Figure 1.6 Integrated photonics elements

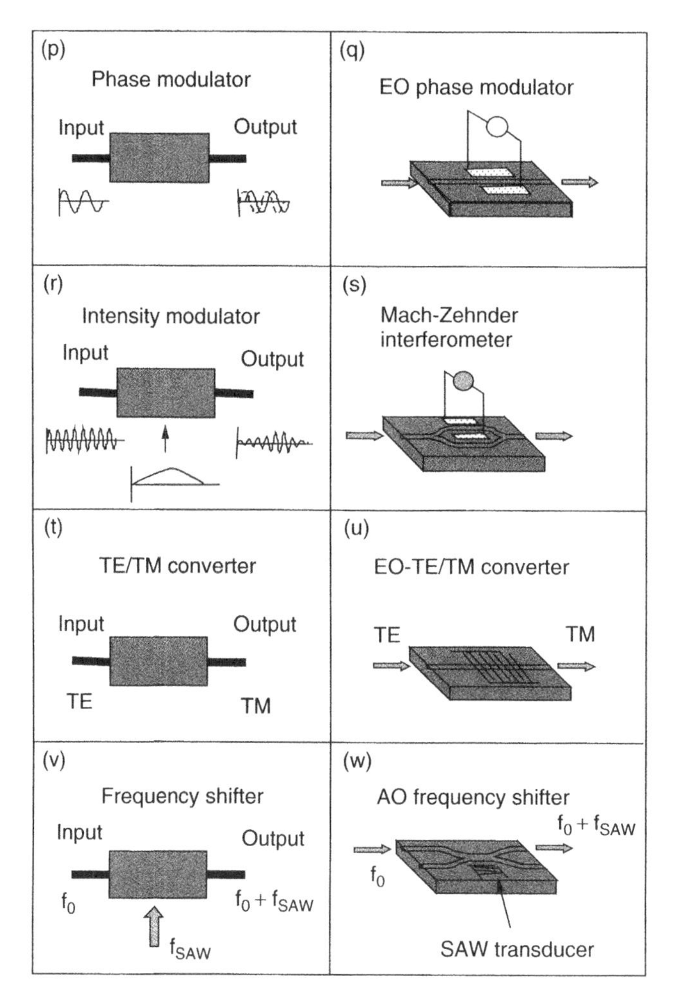

Figure 1.6 *(continued)*

to implement a grating in a region of the straight waveguide (Figure 1.6i). The grating is inherently a wavelength selective element, and thus the grating period must be calculated for the specific working wavelength. The reflection coefficient of the grating depends on the length of the grating region and on the modulation refractive index depth. The wavelength selectivity of the grating is also used for designing waveguide filters working under Bragg condition. Besides this, the grating in integrated photonics can be used as an optical element for performing a wide range of functions such as focusing, deflection, coupling and decoupling light in the waveguide, feedback in an integrated laser, sensors, etc.

- *Directional coupler*. This element has two input ports and two output ports (Figure 1.6j), and is composed of two closely spaced waveguides (Figure 1.6k). The working principle of the coupler is based on the periodical optical power exchange that occurs between two adjacent waveguides through the overlapping of the evanescent waves of the propagating modes. This effect is described by the coupled mode formalism described in Chapter 4. By setting design parameters, including waveguide spacing and coupler length, the ratio of powers between the two output ports may be set during the fabrication process to be between zero and 1.

- *Polariser*. A waveguide polariser allows to pass light having a well defined polarisation character, either TE or TM light, by filtering one of them (Figure 1.6l). The fabrication of a waveguide polariser is as simple as depositing a metallic film onto a waveguide (Figure 1.6m): the light propagating along the waveguide with its electric field perpendicular to the substrate plane (TM mode) is strongly attenuated because of the resonant coupling with the superficial plasmon modes. In this way, at the waveguide output, only light with TE polarisation is present. As the TE mode also suffers some attenuation, the nature of the metal as well as the metallic film length must be carefully chosen in order to obtain a high polarisation ratio, while maintaining a high enough TE light power. An alternative way of obtaining a waveguide polariser is to design a waveguide that supports only TE polarised modes. These are obtained, for example, in lithium niobate waveguides fabricated by the protonic exchange method. In this fabrication process, while the extraordinary index increases, the ordinary index decreases, thus forming a waveguide that supports only extraordinary polarised modes.
- *Polarisation beam splitter*. In some integrated optical devices, it is necessary to divide the input light into its two orthogonal polarisation, TE and TM, in two separate waveguide output ports (Figure 1.6n). Figure 1.6o shows an integrated optical element based on a lithium niobate substrate, which performs this function: the intersecting waveguide operates as a directional coupler whose behaviour depends on the beat between odd-mode light and even-mode light for TE-mode and TM-mode light, respectively [13]. The TE-mode light propagates to the cross-output port and the TM-mode light to the parallel output port. This polarisation selectivity is based on the birefringence of $LiNbO_3$. The length and the width of the intersecting region must be carefully controlled to obtain high extinction ratios of both polarisations, for a chosen wavelength.
- *Phase modulator*. An integrated optical phase modulator performs a controlled shift on the phase of a light beam (Figure 1.6p), and consists of a channel waveguide fabricated on a substrate with the possibility of changing its refractive index by means of an externally applied field (thermal, acoustic, electric, etc.). The most common phase modulator is based on the electro-optic effect: an electric field applied to an electro-optic material, such as $LiNbO_3$, induces a change in its refractive index. If the electric field is applied through a channel waveguide, the change in the refractive index induces a change in the propagation constant of the propagating mode, and therefore the light travelling through that region undergoes a certain phase shift (Figure 1.6q). The geometry of the electrodes and the voltage control depend on the crystal orientation and on the device structure. For high modulation frequency a special electrode configuration is necessary, such as the *travelling wave* configuration or *phase reversal electrodes* configuration.
- *Intensity modulator*. One of the most important functions of an optical chip is the intensity modulation of light at very high frequencies (Figure 1.6r). One of the most simple ways to perform this task is to build an integrated Mach-Zehnder interferometer (MZI) on an electro-optic substrate (Figure 1.6s). The MZI starts with a channel monomode waveguide, and then splits it in two symmetric branches by means of a

Y-branch. After some distance, the two branches becomes parallel. The MZI continues with a symmetric reverse Y-branch, and ends in a straight waveguide. If the MZI is exactly symmetric, the input light splits at the first Y-junction into the two parallel branches, and then recombines constructively into the final straight waveguide. On the contrary, if in one of the interferometer's arms the light suffers a phase shift of 180°, at the end of the second Y-branch the light coming from the two branches will recombine out of phase, and will give rise to destructive interference, with no light at the output. In practice, the phase shift in one arm is carried out via the electro-optic effect, by applying a voltage across the waveguide. By adequately choosing the crystal orientation, polarisation, electrode geometry and applied voltage, a total phase shift of 180° can be obtained for a specific wavelength.

- *TE/TM mode converter*. In a normal situation, TE and TM modes are orthogonal, and then the power transfer between them cannot occur. Nevertheless, TE to TM conversion (Figure 1.6t) can be achieved by using electro-optic substrates, which must have non-zero off-diagonal elements in the electro-optic coefficient matrix. If lithium niobate is used as a substrate, a periodic electrode is required because this crystals is birefringent, and therefore the TE and TM modes have different effective refractive indices (propagation speeds) (Figure 1.6u). By combining phase modulators and a TE/TM converter, a fully integrated polarisation controller can be built.
- *Frequency shifter*. Frequency shifting in integrated optics (Figure 1.6v) can be performed by means of the acousto-optic effect. An acoustic surface wave (SAW) generated by a piezo-electric transducer, creates a Bragg grating in the acousto-optic substrate that interacts with the propagating light in a specially designed region, giving rise to diffracted light that is frequency-shifted by the Doppler effect (Figure 1.6w). This frequency shift corresponds to the frequency of the acoustic wave.

1.6 Some Examples of Integrated Photonics Devices

The optical elements that can be found in an optical chip can be classified according to their function as passive, functional, active and non-linear. A passive optical element has fixed input/output characteristics, which are determined when the photonic component is fabricated. Examples of these are the power splitter, waveguide reflector, directional coupler, polariser, and polarisation beam splitter. Functional optical elements are photonic components which are driven by externally applied fields (for example, electric, acoustic or thermal). The above described phase modulator, intensity modulator, frequency converter and electro-optic TE/TM converter fall into this category. Although some authors call these devices active devices, we will keep the name "active devices" for photonic components that perform functions such as optical amplification and laser oscillation. This choice of nomenclature is due to the fact that they use active impurities such as rare earths embedded in the waveguide structure, to obtain light amplification (or oscillation) via a luminescence process after optical (or electrical) pumping. The integrated optical amplifier and the integrated laser are two examples of active devices. Finally, some integrated optical devices make use of the non-linearity of certain materials to perform frequency doubling or optical parametric oscillation, where the optical chip's function is to generate new frequencies via a non-linear optical process. Since the efficiency of non-linear processes is proportional to the

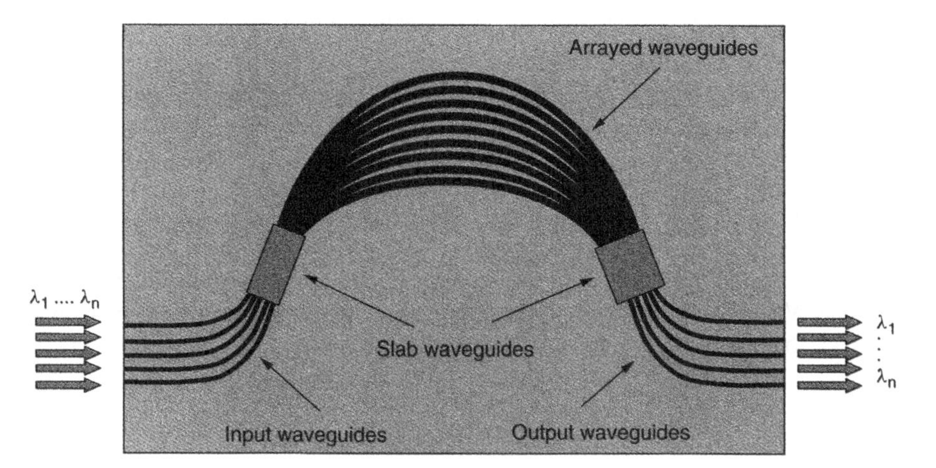

Figure 1.7 The arrayed waveguide grating (AWG) is one example of a passive integrated photonics device, used for dense wavelength demultiplexing

light intensity, these devices yield a very good performance in the integrated photonic version, because of the small transverse area of the waveguide propagating beams.

Figure 1.7 shows an example of a passive integrated photonic device, in which no external signal is needed for its operation. This device is called an arrayed waveguide grating (AWG, PHASAR or waveguide grating router, WGR); its function is to passively multiplex or demultiplex signals of closely spaced wavelengths, and it is used in fibre optical communication systems [14]. Several wavelengths coming through a single fibre enter the AWG via any of its input waveguides. A coupler splits light between many of the curved waveguides which define the AWG. The arrayed waveguides are formed by waveguides having different lengths, and therefore light suffers different phase shift for each curved waveguide. By precisely adjusting the phase shift from each curved waveguide with respect to all the others, an interferometric pattern is set up that results in light of different wavelengths being focused at different spatial location on an output arc. Since the AWG distribute signals according to their wavelength, each individual waveguide output corresponds to a specific wavelength, thereby acting as a demultiplexor.

An example of a functional device, which also combines some passive elements is the acousto-optic tuneable filter (AOTF) (Figure 1.8) [15]. This integrated optical device requires an external radio-frequency (RF) control signal to selectively separate one or more wavelength signals (drop signals). This device is fabricated with $LiNbO_3$ and is composed of a piezo-transducer, a thin film acoustic waveguide and two polarisation beam splitters. The multi-wavelength input signals propagate over the optical waveguide and are divided into their perpendicular components (TE/TM) by the first polarisation beam splitter (PBS). Surface acoustic waves (SAW), generated by applying an RF signal to the transducer, travel through the SAW guide and cause a periodic modulation of the optical waveguide's refractive index. The periodic refractive index change induces TE–TM or TM–TE conversion for the drop wavelength only. The drop wavelength corresponds to the applied RF frequency and becomes perpendicular to the incident light. The second PBS is then used to separate the drop wavelength from the incident light. By using several RF signals simultaneously, it is even possible to drop several wavelengths.

Several substrate materials compatible with integrated photonic technology are also suitable to incorporate optically active rare earth ions, which makes it possible to fabricate active integrated optical devices [16]. Figure 1.9 shows the arrangement of an integrated optical amplifier based on Erbium and Ytterbium ions. It basically consists of a straight waveguide, which has rare earth ions incorporated to it, an undoped waveguide and a directional coupler. The input pumping at 980 nm is injected into the undoped waveguide, and the coupler transfers the pump energy to the doped straight waveguide. Via several radiative, non-radiative and energy transfer mechanisms which takes place on the Erbium and Ytterbium ions, the feeble input signal at 1533 nm

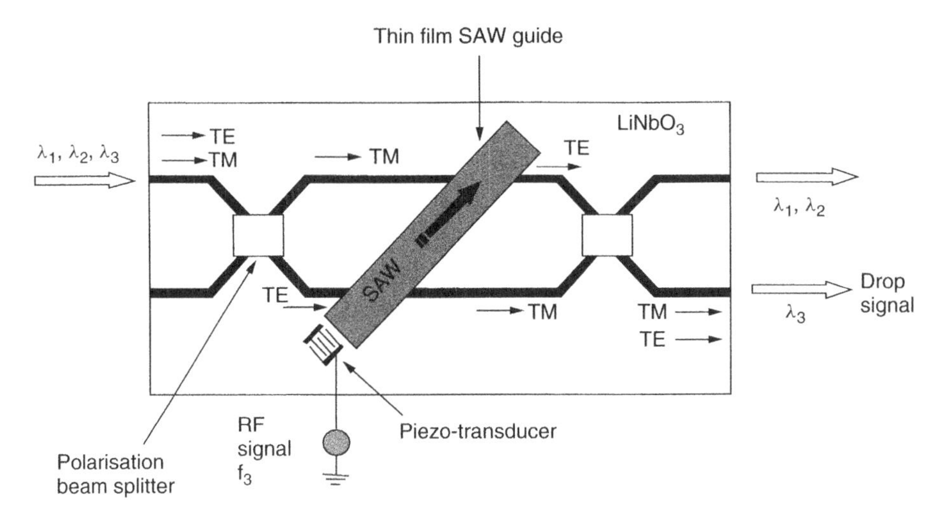

Figure 1.8 The integrated acousto-optic tuneable filter uses polarisation conversion, via the interaction between the light guide modes and the surface acoustic waves generated by a piezo-electric transducer, to spatially separate any of the selected input wavelengths (drop signals). Since the device is externally controlled by the RF frequency applied to the transducer, this is one example of a functional integrated photonic device

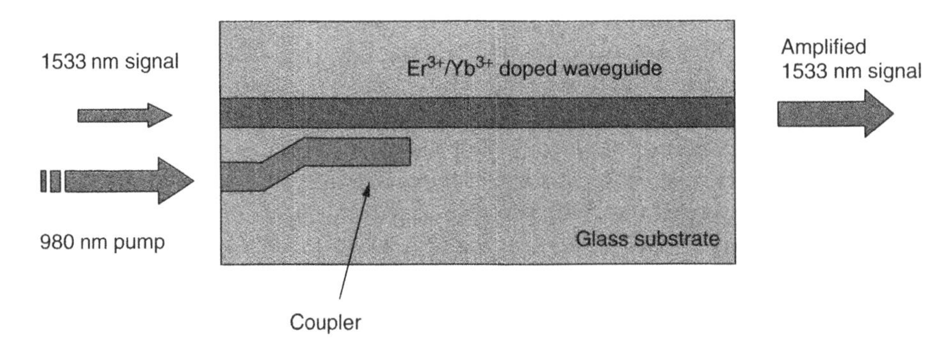

Figure 1.9 The integrated optical amplifier based on rare earths is one example of active integrated photonic chips

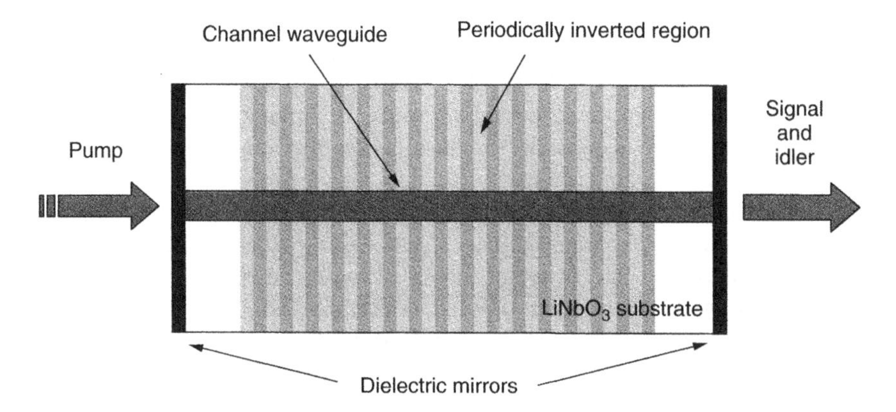

Figure 1.10 The high optical non-linear coefficients of $LiNbO_3$ crystals make this substrate suitable for developing non-linear integrated photonic devices, such as the optical parametric oscillator presented here. Also, for high efficiency conversion, it is necessary to fabricate a periodically poled region along the waveguide structure

is amplified as it propagates along the straight waveguide. If a couple of dielectric mirrors are attached at the two waveguide ends, the amplified signal can oscillate, and therefore an integrated laser can be obtained. The end mirrors can also be replaced by integrated gratings, acting as a true wavelength-selective reflector.

Integrated optical parametric oscillators (OPOs) in ferro-electric crystals have been identified as the most useful tuneable non-linear frequency converters with many applications, mainly in environmental sensing and process monitoring. These non-linear integrated photonic devices are based on ferro-electric materials showing high values of second order nonlinearities, and are capable of obtaining a periodic inversion of the ferroelectric domains. Figure 1.10 presents the design of an optical parametric oscillator in its integrated optical version: a straight channel waveguide is fabricated on a z-cut $LiNbO_3$ substrate, where a periodically poled region has been patterned perpendicular to the waveguide [17]. The two dielectric mirrors, directly attached to the waveguide ends, allow parametric oscillation at the signal and idler frequencies, which are created from the input pump via non-linear optical interactions. For efficient optical parametric oscillation the crystal orientation must be adequately chosen, as well as the periodicity of the ferro-electric domain structure.

1.7 Structure of the Book

The rest of the book has been divided into four chapters and some appendices. This first chapter aimed to present an overview of integrated photonic technology, stressing the radical conceptual change of photonic chips compared to traditional optical systems. Although several technical terms have been used throughout this chapter (modes, coupling, TE/TM conversion, etc.) without a rigorous definition, they will be further studied in subsequent chapters. Chapter 2 gives the basic EM theory necessary for developing and understanding light behaviour in waveguide structures, starting from Maxwell's equations. The theory of optical waveguides is introduced in Chapter 3.

For a correct description of light in waveguide structures having dimensions comparable to its wavelength, the light must be contemplated as EM waves. Therefore, the waveguide theory discussed in Chapter 3 is based on the EM theory of light, where the important concept of optical waveguide mode is introduced. In this chapter we start analysing the planar waveguide structure, where the most relevant concepts are explained. Also, once one-dimensional waveguides are studied (planar waveguides), we focus our attention on the theory of guided modes in two-dimensional structures such as channel waveguides, which are the basic elements in photonic integrated circuits.

Chapter 4 is devoted to the coupling theory of modes in optical waveguides. The understanding of mode coupling is of vital importance for most integrated optical devices. This chapter includes the study of optical power transfer between waveguide modes, whether it is energy transfer between co-directional or contradirectional propagating modes. Also, waveguide diffraction gratings are introduced in this chapter, as they are key integrated photonic elements which offer an efficient and controllable way of exchange power between waveguide modes.

Finally, Chapter 5 deals with the theory of light propagation in waveguide structures. The problem of optical propagation in waveguides is reducible to solve light paraxial propagation in inhomogeneous media, where paraxial means propagation mainly along a preferential direction. Although we will discuss several approaches to this problem, we will focus on the beam propagation algorithm, known as beam propagation method (BPM), which is a step-by-step method of simulating the passing of light through any waveguiding medium, allowing us to track the optical field at any point as it propagates along guiding structures.

References

[1] T.H. Maiman, "Stimulated Optical Radiation in Ruby Masers", *Nature*, **187**, 493–494 (1960).

[2] S.E. Miller, "Integrated Optics: An Introduction", *The Bell System Technical Journal*, **48**, 2059–2068 (1969).

[3] N. Hairston, "Lithography Aids Integration of Optics", *Laser Focus World*, **37**, November 2001, 93–98.

[4] M.A. Powell and A.O. Donnell, "What Integrated Optics is Really Used For", *Optics and Photonics News*, September 1997, 23–29.

[5] L. Gasman, "New Materials Renew Life for Integrated Optics", *WDM Solutions*, November 2001, 17–20.

[6] H. Nishihara, M. Haruna and T. Suhara, *Optical Integrated Circuits*. McGraw-Hill, New York (1985).

[7] S.I. Najafi, *Introduction to Glass Integrated Optics*, Artech House, Boston (1992).

[8] A.M. Prokhorov, Y.S. Kuz'minov and O.A. Khachaturyan, *Ferroelectric Thin-Film Waveguides in Integrated Optics and Optoelectronics*. Cambridge International Science Publishing (1996).

[9] M. Kawachi, "Silica Waveguides on Silicon and their Application to Integrated-Optic Components", *Optical and Quantum Electronics*, **22**, 391–416 (1990).

[10] H. Volterra and M. Zimmerman, "Indium Phosphide Benefits High-Performance Transmission", *WDM Solutions*, October 2000, 47–49.

[11] K. Wakao, H. Soda and Y. Kotaki, "Semiconductor Optical Active Devices for Photonics Networks", *FUJITSU Scientific and Technical Journal*, **35**, 100–106 (1999).

[12] M. Cowin, "Manufacturers Poised to Profit from Polymeric Breakthroughs", *Fibre Systems Europe*, October 2001, 113–118.
[13] T. Nakazawa, S. Taniguchi and M. Seino, "Ti:LiNbO$_3$ Acousto-Optic Tunable Filter (AOTF)", *FUJITSU Scientific and Technical Journal*, **35**, 107–112 (1999).
[14] M.K. Smit and C. van Dam, "Phasar Based WDM Devices: Principles, Design, and Applications", *IEEE J. Select Topics in Quantum Electronics*, **2**, 236–250 (1996).
[15] T. Chikama, H. Onaka and S. Kuroyanagi, "Photonic Networking Using Optical Add Drop Multiplexers and Optical Cross-Connects", *FUJITSU Scientific and Technical Journal*, **35**, 46–55 (1999).
[16] J. Hübner, S. Guldberg-Kjaer, M. Dyngaard, Y. Shen, C.L. Thomsen, S. Balslev, C. Jensen, D. Zauner and T. Feuchter, "Planar Er- and Yb-Doped Amplifiers and Lasers", *Applied Physics B*, **73**, 435–438 (2001).
[17] D. Hofmann, H. Herrmann, G. Schreiber, W. Grundkötter, R. Ricken and W. Sohler, "Continuous-Wave Mid-Infrared Optical Parametric Oscillators with Periodically Poled Ti:LiNbO$_3$ Waveguide", *Proc. Europ. Conf. on Integrated Optics* (ECIO'99), Turin/Italy, April 1999, post-deadline paper, 21–24.

Further Reading

Applied Physics B, Lasers and Optics **73**, N. 5–6 (2001), special issue on Integrated Optics.
Integrated Optical Circuits and Components: Design and Application, ed. by E.J. Murphy, Marcel Dekker, New York, 1999.

2

REVIEW OF THE ELECTROMAGNETIC THEORY OF LIGHT

Introduction

As we saw in the introductory chapter, integrated photonics devices are based on optical waveguides with depth of the order of microns, comparable to the wavelength of the optical radiation used in these devices (visible and near infrared). This fact implies that the performance of the optical chips cannot be analysed in terms of ray optics, but instead the light must be treated as electromagnetic waves. Therefore, the electromagnetic theory of light is necessary to properly describe the behaviour of the different optical elements that are present in any integrated photonics device. In some cases, the vectorial nature of the electromagnetic waves can be simplified, and a scalar treatment of the optical waves is often enough for a reasonable description of the phenomena involved.

In this chapter we present the basics of the electromagnetic theory of light. We derive the wave equation starting from Maxwell's equations for light propagation in free space, and then the wave equation in dielectric media is obtained, by introducing the definition of refractive index. The solution for the temporal part of the wave equation admits solution in the form of harmonic functions, which is then used to derive a wave equation for monochromatic waves, where only the spatial dependence of the electromagnetic field needs to be considered. The so-called Helmholtz equation is indeed the starting equation for the analysis of optical waveguides. We then study the properties of plane waves, as a particular solution of the Helmholtz equation, and describe the behaviour of electromagnetic waves from the point of view of the vectorial nature of the electric and magnetic fields, in terms of their polarisation character. Losses in passive waveguides, as well as gain in active waveguides (lasers and amplifiers) are also important topics when dealing with light propagation in optical waveguide structures. To describe the behaviour of electromagnetic radiation in absorbing/gain media in a general manner, we derive a more general wave equation by defining a complex refractive index.

Optical waveguides are inherently inhomogeneous structures, in the sense that different media with different optical constants are necessary to achieve light confinement. The present chapter deals with the behaviour of light at dielectric interfaces, and the reflected and transmitted waves are described by defining the reflection and transmission coefficients, where the two types of incident waves (TE and TM polarised waves) are examined separately. Also, the energy relations between incident, reflected and transmitted waves are derived. Finally, the important phenomenon of total internal reflection, being a key topic in the understanding and description of optical waveguides, is discussed.

2.1 Electromagnetic Waves

2.1.1 Maxwell's equations: wave equation

Light is, according to classical theory, the flow of electromagnetic (EM) radiation through free space or through a medium in the form of electric and magnetic fields. Although electromagnetic radiation covers an extremely wide range, from gamma rays to long radio waves, the term "light" is restricted to the part of the electromagnetic spectrum that goes from the vacuum ultraviolet to the far infrared. This part of the spectrum is also called *optical range*. EM radiation propagates in the form of two mutually perpendicular and coupled vectorial waves: the electric field $\mathcal{E}(\mathbf{r}, t)$ and the magnetic field $\mathcal{H}(\mathbf{r}, t)$. These two vectorial magnitudes depend on the position ($\mathbf{r}$) and time (t). Therefore, in order to properly describe light propagation in a medium, whether vacuum or a material, it is necessary in general to know six scalar functions, with their dependence of the position and the time. Fortunately, these functions are not completely independent, because they must satisfy a set of coupled equations, known as Maxwell's equations.

Maxwell's equations form a set of four coupled equations involving the electric field vector and the magnetic field vector of the light, and are based on experimental evidence. Two of them are scalar equations, and the other two are vectorial. In their differential form, Maxwell's equations for light propagating in free space are:

$$\nabla \mathcal{E} = 0 \tag{2.1}$$

$$\nabla \mathcal{H} = 0 \tag{2.2}$$

$$\nabla \times \mathcal{E} = -\mu_0 \frac{\partial \mathcal{H}}{\partial t} \tag{2.3}$$

$$\nabla \times \mathcal{H} = \varepsilon_0 \frac{\partial \mathcal{E}}{\partial t} \tag{2.4}$$

where the constants $\varepsilon_0 = 8.85 \times 10^{-12}$ m^{-3} kg^{-1} s^4 A^2 and $\mu_0 = 4\pi \times 10^{-7}$ mkgs^{-2} A^{-2} represent the *dielectric permittivity* and the *magnetic permeability* of free space respectively, and the ∇ and ∇x denote the divergence and curl operators, respectively.

For the description of the electromagnetic field in a material medium it is necessary to define two additional vectorial magnitudes: the *electric displacement vector* $\mathcal{D}(\mathbf{r}, t)$ and the *magnetic flux density vector* $\mathcal{B}(\mathbf{r}, t)$. Maxwell's equations in a material medium, involving these two magnitudes and the electric and magnetic fields, are expressed as:

$$\nabla \mathcal{D} = \rho \tag{2.5}$$

$$\nabla \mathcal{B} = 0 \tag{2.6}$$

$$\nabla \times \mathcal{E} = -\frac{\partial \mathcal{B}}{\partial t} \tag{2.7}$$

$$\nabla \times \mathcal{H} = \mathcal{J} + \frac{\partial \mathcal{D}}{\partial t} \tag{2.8}$$

where $\rho(\mathbf{r}, t)$ and $\mathcal{J}(\mathbf{r}, t)$ denote the *charge density* and the *current density vector* respectively. If in the medium there are no free electric charges, which is the most common situation in optics, Maxwell's equations simplify in the form:

$$\nabla \mathcal{D} = 0 \tag{2.9}$$

$$\nabla \mathcal{B} = 0 \tag{2.10}$$

$$\nabla \times \mathcal{E} = -\frac{\partial \mathcal{B}}{\partial t} \tag{2.11}$$

$$\nabla \times \mathcal{H} = \mathcal{J} + \frac{\partial \mathcal{D}}{\partial t} \tag{2.12}$$

Nevertheless, to solve these differential coupled equations it is necessary to establish additional relations between the vectors $\mathcal{D}$ and $\mathcal{E}$, $\mathcal{J}$ and $\mathcal{E}$, and the vectors $\mathcal{H}$ and $\mathcal{B}$. These relations are called *constitutive relations*, and depend on the electric and magnetic properties of the considered medium. For a *linear*, *homogeneous* and *isotropic* medium, the constitutive relations are given by:

$$\mathcal{D} = \varepsilon \mathcal{E} \tag{2.13}$$

$$\mathcal{B} = \mu \mathcal{H} \tag{2.14}$$

$$\mathcal{J} = \sigma \mathcal{E} \tag{2.15}$$

where ε is the dielectric permittivity, μ is the magnetic permeability and σ is the conductivity of the medium. If the medium is not linear, it should be necessary to include additional terms involving power expansion of the electric and magnetic fields. On the other hand, the fact of assuming a homogeneous medium implies that the optical constants of the medium ε, μ and σ are not dependent of the position vector $\mathbf{r}$. Finally, in an isotropic medium these optical constants are scalar magnitudes and independent of the direction of the vectors $\mathcal{E}$ and $\mathcal{H}$, implying that the vectors $\mathcal{D}$ and $\mathcal{J}$ are parallel to the electric field $\mathcal{E}$, and the vector $\mathcal{B}$ is parallel to the magnetic field $\mathcal{H}$. By contrast, in an anisotropic medium the optical constants must be treated as tensorial magnitudes, and the above mentioned parallelism is no longer valid in general.

By using the constitutive relations for a linear, homogeneous and isotropic medium, Maxwell's equations can be written in terms of the electric field $\mathcal{E}$ and magnetic field $\mathcal{H}$ only:

$$\nabla \mathcal{E} = 0 \tag{2.16}$$

$$\nabla \mathcal{H} = 0 \tag{2.17}$$

$$\nabla \times \mathcal{E} = -\mu \frac{\partial \mathcal{H}}{\partial t} \tag{2.18}$$

$$\nabla \times \mathcal{H} = \sigma \mathcal{E} + \varepsilon \frac{\partial \mathcal{E}}{\partial t} \tag{2.19}$$

By combining adequately these four differential equations, it is possible to obtain two differential equations in partial derivatives, one for the electric field and another for the magnetic field:

$$\nabla^2 \boldsymbol{\mathcal{E}} = \mu\sigma \frac{\partial \boldsymbol{\mathcal{E}}}{\partial t} + \mu\varepsilon \frac{\partial^2 \boldsymbol{\mathcal{E}}}{\partial t^2} \tag{2.20}$$

$$\nabla^2 \boldsymbol{\mathcal{H}} = \mu\sigma \frac{\partial \boldsymbol{\mathcal{H}}}{\partial t} + \mu\varepsilon \frac{\partial^2 \boldsymbol{\mathcal{H}}}{\partial t^2} \tag{2.21}$$

These two differential equations are known as *wave equations* for a material medium. It is worth noting that, although we have obtained a wave equation for the electric field $\boldsymbol{\mathcal{E}}$ and another for the magnetic field $\boldsymbol{\mathcal{H}}$, the solution of both equations are not independent, because the electric and magnetic fields are related through Maxwell's equations (2.18) and (2.19).

2.1.2 Wave equation in dielectric media

A perfect dielectric medium is defined as a material in which the conductivity is $\sigma = 0$. In this category fall most of the substrate materials used for integrated optical devices, such as glasses, ferro-electric crystals or polymers, while metals do not belong to this category because of their high conductivity. Then, for dielectric media ($\sigma = 0$) the wave equations simplify on the forms:

$$\nabla^2 \boldsymbol{\mathcal{E}} = \mu\varepsilon \frac{\partial^2 \boldsymbol{\mathcal{E}}}{\partial t^2} \tag{2.22}$$

$$\nabla^2 \boldsymbol{\mathcal{H}} = \mu\varepsilon \frac{\partial^2 \boldsymbol{\mathcal{H}}}{\partial t^2} \tag{2.23}$$

Each of these two vectorial wave equations can be separated on three scalar wave equations, expressed as:

$$\nabla^2 \xi = \mu\varepsilon \frac{\partial^2 \xi}{\partial t^2} \tag{2.24}$$

where the scalar variable $\xi(\mathbf{r}, t)$ may represent each of the six Cartesian components of either the electric and magnetic fields. The solution of this equation represents a wave that propagates with a speed v (*phase velocity*) given by:

$$v = \frac{1}{\sqrt{\varepsilon\mu}} \tag{2.25}$$

Therefore, the complete solution of the vectorial wave equations (2.22) and (2.23) represents an *electromagnetic wave*, where each of the Cartesian components of the electric and magnetic fields propagate in the form of waves of equal speed v.

For propagation in free space, and using the values for ε_0 and μ_0 we obtain:

$$c = \frac{1}{\sqrt{\varepsilon_0\mu_0}} \approx 3.00 \times 10^8 \text{ m/s} \tag{2.26}$$

which corresponds to the speed of light in free space measured experimentally. It is worth noting that here the speed of light has been obtained only using values of electric and magnetic constants.

Usually it is convenient to express the propagation speed of the electromagnetic waves in a medium v as function of the speed of light in free space c, through the relation:

$$v \equiv \frac{c}{n} \tag{2.27}$$

where n represents the *refractive index* of the dielectric medium. Taking into account the relations (2.25), (2.26) and (2.27), the refractive index is related with the optical constant of the material medium and the dielectric permittivity and the magnetic permeability of the free space by:

$$n = \sqrt{\frac{\varepsilon\mu}{\varepsilon_0\mu_0}} \tag{2.28}$$

As we will see in the following chapters, the refractive index of a medium is the most important parameter in integrated photonic technology.

In most of the materials (non-magnetic materials), and in particular in dielectric media, the magnetic permeability is very close to that of free space: $\mu \approx \mu_0$. With this approximation (a very good one, indeed), the refractive index can be simplified to obtain:

$$n \approx \sqrt{\frac{\varepsilon}{\varepsilon_0}} = \sqrt{\varepsilon_r} \tag{2.29}$$

where we have introduced the magnitude *relative dielectric permittivity* ε_r, also often called *dielectric constant*, defined as the relation between the dielectric permittivity of the material medium and that of the free space. Table 2.1 summarises the refractive indices of the most relevant materials used in integrated photonic technology. Besides the refractive index of 1 corresponding to propagation through the free space, as can be seen in the Table 2.1 the refractive index ranges from values close to 1.5 for glasses and some dielectric crystals to values close to 4 for semiconductor materials.

Table 2.1 Refractive indices corresponding to materials commonly used in the fabrication of integrated photonic components

Material	Refractive index	Wavelength (nm)
Glass (BK7)	1.51	633
Glass (ZBLAN)	1.50	633
Polymer (PMMA)	1.54	633
Silica (amorphous SiO_2)	1.45	633
Quartz (SiO_2)	1.55	633
Silicon nitride (Si_3N_4)	2.10	633
Calcium fluoride (CaF_2)	1.43	633
Lithium niobate ($LiNbO_3$)	2.28 (n_o) 2.20 (n_e)	633
Silicon (Si)	3.75	1300
Gallium arsenide (GaAs)	3.4	1000
Indium phosphide (InP)	3.17	1510

The electromagnetic waves transport energy, and the flux of energy carried by the EM wave is given by the *Poynting vector* $\mathcal{S}$, defined as:

$$\mathcal{S} \equiv \mathcal{E} \times \mathcal{H} \tag{2.30}$$

On the other hand, the *intensity* (or *irradiance*) I of the EM wave, defined as the amount of energy passing through the unit area in the unit of time, is given by the time average of the Poynting vector modulus:

$$I = \langle |\mathcal{S}| \rangle \tag{2.31}$$

The fact of using an averaged value instead of an instant value to define the intensity of an EM wave is because, as we will see in the next section, the electric and magnetic fields associated with the EM wave oscillate at very high frequency, and the apparatus used to detect that intensity (light detectors) cannot follow the instant values of the Poynting vector modulus.

2.1.3 Monochromatic waves

The time dependence of the electric and magnetic fields within the wave equations admits solutions of the form of harmonic functions. Electromagnetic waves with such sinusoidal dependence on the time variable are called *monochromatic waves*, and are characterised by their *angular frequency* ω. In a general form, the electric and magnetic fields associated with a monochromatic wave can be expressed as:

$$\mathcal{E}(\mathbf{r}, t) = \mathcal{E}_0(\mathbf{r}) \cos[\omega t + \varphi(\mathbf{r})] \tag{2.32}$$

$$\mathcal{H}(\mathbf{r}, t) = \mathcal{H}_0(\mathbf{r}) \cos[\omega t + \varphi(\mathbf{r})] \tag{2.33}$$

where the *fields amplitudes* $\mathcal{E}_0(\mathbf{r})$ and $\mathcal{H}_0(\mathbf{r})$ and the initial phase $\varphi(\mathbf{r})$ depend on the position $\mathbf{r}$, but the time dependence is carried out only in the cosine argument through ωt.

When dealing with monochromatic waves, in general it is easier to write down the monochromatic fields using *complex notation*. Using this notation, the electric and magnetic fields are expressed as:

$$\mathcal{E}(\mathbf{r}, t) = Re[\mathbf{E}(\mathbf{r})e^{+i\omega t}] \tag{2.34}$$

$$\mathcal{H}(\mathbf{r}, t) = Re[\mathbf{H}(\mathbf{r})e^{+i\omega t}] \tag{2.35}$$

where $\mathbf{E}(\mathbf{r})$ and $\mathbf{H}(\mathbf{r})$ denote the *complex amplitudes* of the electric and magnetic fields, respectively (see Appendix 1). The *angular frequency* ω that characterises the monochromatic wave is related to the *frequency* ν and the *period* T by:

$$\omega = 2\pi\nu = 2\pi/T \tag{2.36}$$

The electromagnetic spectrum covered by light (optical spectrum) ranges from frequencies of 3×10^5 Hz corresponding to the far IR, to 6×10^{15} Hz corresponding to vacuum UV, being the frequency of visible light around 5×10^{14} Hz.

The average of the Poynting vector as a function of the complex fields amplitudes for monochromatic waves takes the form:

$$\langle \mathcal{S} \rangle = \langle Re\{\mathbf{E}e^{+i\omega t}\} \times Re\{\mathbf{H}e^{+i\omega t}\}\rangle = Re\{\mathbf{S}\} \tag{2.37}$$

where $\mathbf{S}$ has been defined as:

$$\mathbf{S} = 1/2\ \mathbf{E} \times \mathbf{H}^* \tag{2.38}$$

and is called the *complex Poynting vector.* In this way, the intensity carried by a monochromatic EM wave should be expressed as:

$$I = |Re\{\mathbf{S}\}| \tag{2.39}$$

In the case of monochromatic waves, Maxwell's equations using the complex fields amplitudes $\mathbf{E}$ and $\mathbf{H}$ are simplified notably, because the partial derivatives in respect of the time are directly obtained by multiplying by the factor $i\omega$:

$$\nabla\mathbf{E} = 0 \tag{2.40}$$

$$\nabla\mathbf{H} = 0 \tag{2.41}$$

$$\nabla \times \mathbf{E} = -i\mu_0\omega\mathbf{H} \tag{2.42}$$

$$\nabla \times \mathbf{H} = i\varepsilon\omega\mathbf{E} \tag{2.43}$$

where we have assumed a dielectric and non-magnetic medium in which $\sigma = 0$ and $\mu = \mu_0$.

Now, if we substitute the solutions on the form of monochromatic waves (2.34) and (2.35) in the wave equation (2.24), we obtain a new wave equation, valid only for monochromatic waves, known as the *Helmholtz equation*:

$$\nabla^2 U(\mathbf{r}) + k^2 U(\mathbf{r}) = 0 \tag{2.44}$$

where now $U(\mathbf{r})$ represents each of the six Cartesian components of the $\mathbf{E}(\mathbf{r})$ and $\mathbf{H}(\mathbf{r})$ vectors defined in (2.34) and (2.35), and where we have defined k as:

$$k \equiv \omega(\varepsilon\mu_0)^{1/2} = nk_0 \tag{2.45}$$

$$k_0 \equiv \omega/c \tag{2.46}$$

If the material medium is inhomogeneous the dielectric permittivity is no longer constant, but position dependent $\varepsilon = \varepsilon(\mathbf{r})$. In this case, although Maxwell's equations remain valid, the wave equation (2.24) or the Helmholtz equation (2.44) are not longer valid. Nevertheless, for a *locally homogeneous medium*, in which $\varepsilon(\mathbf{r})$ varies slowly for distances of $\sim 1/k$, those wave equations are approximately valid by now defining $k = n(\mathbf{r})k_0$, and $n(\mathbf{r}) = [\varepsilon(\mathbf{r})/\varepsilon_0)]^{1/2}$.

2.1.4 Monochromatic plane waves in dielectric media

Once the temporal dependence of the electromagnetic fields has been established in terms of monochromatic waves, let us now consider the spatial dependence of the fields. For monochromatic waves, the solution for the spatial dependence, carried by the complex amplitudes $\mathbf{E}(\mathbf{r})$ and $\mathbf{H}(\mathbf{r})$, can be obtained by solving the Helmholtz equation (2.44). One of the easiest and most intuitive solutions for this equation, and also the most frequently used in optics, is the *plane wave*. The plane wave is characterised by its *wavevector* **k**, and the mathematical expressions for the complex amplitudes are:

$$\mathbf{E}(\mathbf{r}) = \mathbf{E}_0 e^{-i\mathbf{k}\mathbf{r}} \tag{2.47}$$

$$\mathbf{H}(\mathbf{r}) = \mathbf{H}_0 e^{-i\mathbf{k}\mathbf{r}} \tag{2.48}$$

where the magnitudes $\mathbf{E}_0$ and $\mathbf{H}_0$ are now constant vectors. Each of the Cartesian components of the complex amplitudes $\mathbf{E}(\mathbf{r})$ and $\mathbf{H}(\mathbf{r})$ will satisfy the Helmholtz equation, providing that the modulus of the wavevector **k** is:

$$k = nk_0 = (\omega/c)n \tag{2.49}$$

where ω is the angular frequency of the EM plane wave and n is the refractive index of the medium where the wave propagates.

As the solution given by the electric and magnetic complex amplitudes must satisfy Maxwell's equation, by substituting equations (2.47) and (2.48) into (2.42) and (2.43) the following relations are straightforwardly obtained:

$$\mathbf{k} \times \mathbf{H}_0 = -\omega\varepsilon\mathbf{E}_0 \tag{2.50}$$

$$\mathbf{k} \times \mathbf{E}_0 = \omega\mu_0\mathbf{H}_0 \tag{2.51}$$

These two formulae, valid only for plane monochromatic waves, establish the relationship between the electric field **E**, the magnetic field **H** and the wavevector **k** of the plane wave. From equation (2.50) one obtains that the electric field is perpendicular to the magnetic field and the wavevector. In the same way, the relation (2.51) establishes that the magnetic field is perpendicular to **E** and **k**. Therefore, one can conclude that **k**, **E** and **H** are mutually orthogonal, and because **E** and **H** lie on a plane normal to the propagation direction defined by **k**, such wave in called a *transverse EM wave* (TEM) (Figure 2.1).

The fact that these three vectors are perpendicular implies (from equations (2.50) and (2.51)) that $\mathbf{H}_0 = (\omega\varepsilon/k)\mathbf{E}_0$ and $\mathbf{H}_0 = (k/\omega\mu_0)\mathbf{E}_0$. These two relations can be

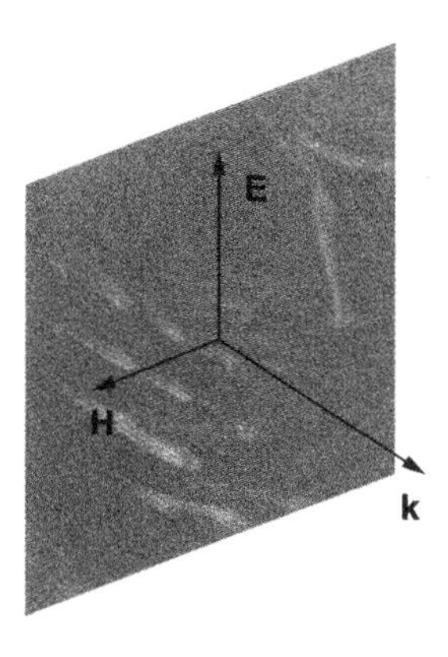

Figure 2.1 Triad defined by the electric field, magnetic field and wavevector, for a plane wave

simultaneously fulfilled only if the wavevector modulus is $k = \omega(\varepsilon\mu_0)^{1/2} = \omega/v = nk_0$. Of course, this is the condition needed for the wave solution described by (2.47) and (2.48) to fulfil the Helmholtz equation (2.44).

When dealing with a monochromatic plane EM wave it is useful to characterise it by its radiation *wavelength* λ, defined as the distance between the two nearest points with equal phase of vibration, measured along the propagation direction. The wavelength is therefore expressed by:

$$\lambda \equiv vT = v/\nu = 2\pi/k = 2\pi/nk_0 = \lambda_0/n \tag{2.52}$$

where λ_0 represents the wavelength of the EM wave in free space, given by:

$$\lambda_0 = cT = c/\nu = 2\pi/k_0 \tag{2.53}$$

It is worth remarking that when an EM wave passes from one medium to another its frequency remains unchanged, but as its phase velocity is modified due to its dependence on the refractive index, the wavelength associated with the EM wave should also change. Therefore, when the wavelength of an EM wave is given, it is usually referred to the wavelength of that radiation propagating through free space.

2.1.5 Polarisation of electromagnetic waves

As we have seen, the electric and magnetic fields of a plane EM wave oscillate in a direction perpendicular to the direction of propagation defined by the wavevector $\mathbf{k}$, and therefore the plane EM waves are *transversal*. A property associated with a transversal wave is its polarisation character, related to the closed curve described by the tip of the electric (or magnetic) field vector at a fixed point $\mathbf{r} = \mathbf{r}_0$ in the space. In order to analyse the polarisation character of an EM plane wave, let us assume, without loss of generality, that the EM wave propagates along the z-axis. In this case we have:

$$\mathbf{k} = k\mathbf{u}_z \tag{2.54}$$

where we will use $\mathbf{u}_x$, $\mathbf{u}_y$ and $\mathbf{u}_z$ as the unitary vectors along the x, y and z-axis respectively.

First, let us assume the simplest situation of an EM wave in which its associated electric field is along the x-axis:

$$\boldsymbol{\mathcal{E}} = E_0 \cos(\omega t - kz)\mathbf{u}_x \tag{2.55}$$

The magnetic field of this EM wave is obtained using the relation (2.51):

$$\boldsymbol{\mathcal{H}} = H_0 \cos(\omega t - kz)\mathbf{u}_y \tag{2.56}$$

where the amplitude H_0 is related to the amplitude E_0 by:

$$H_0 = (k/\omega\mu_0)E_0 = (\varepsilon/\mu_0)^{1/2}E_0 \tag{2.57}$$

Note that the electric and magnetic fields are in phase, that is, if at a fixed time and at a particular plane $z = z_0$ (z being arbitrary) the electric field $\boldsymbol{\mathcal{E}}$ reaches its maximum

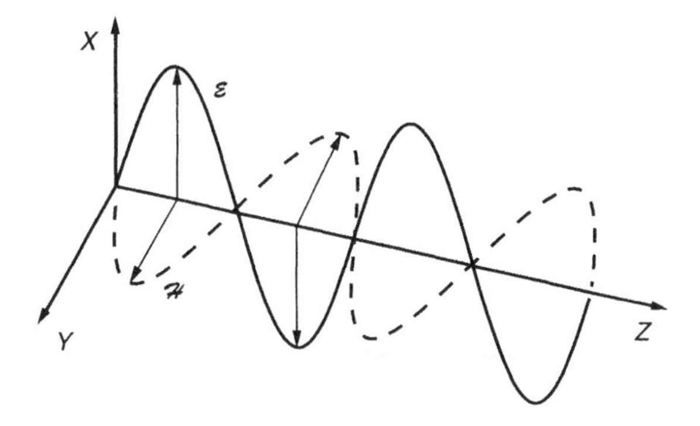

Figure 2.2 Oscillation of the electric and magnetic field vectors on a z-propagating EM plane wave, which is linearly polarised along the x-axis

value, the magnetic field $\mathcal{H}$ will also be at its maximum value. The wave described by equations (2.55) and (2.56) is said to be *linearly polarised* (or more specifically, *linearly x-polarised*) because the electric field vector $\mathcal{E}$ (or $\mathcal{H}$) is always along a particular direction (x direction in this case) (see Figure 2.2).

Let consider now a linearly y-polarised wave, with an addition phase of $+\pi/2$ described by:

$$\mathcal{E} = E_0 \cos(\omega t - kz + \pi/2)\mathbf{u}_y = -E_0 \sin(\omega t - kz)\mathbf{u}_y \tag{2.58}$$

and

$$\mathcal{H} = -H_0 \cos(\omega t - kz + \pi/2)\mathbf{u}_x = H_0 \sin(\omega t - kz)\mathbf{u}_x \tag{2.59}$$

with $H_0 = (\varepsilon/\mu_0)^{1/2} E_0$. Because Maxwell's equations are linear, a linear combination of several solutions will also be a solution. In particular, the sum of plane waves described in (2.55) and (2.56), and those described by (2.58) and (2.59) will give us a different, but valid, solution of the wave equation:

$$\mathcal{E} = E_0[\cos(\omega t - kz)\mathbf{u}_x - \sin(\omega t - kz)\mathbf{u}_y] \tag{2.60}$$

and

$$\mathcal{H} = H_0[\cos(\omega t - kz)\mathbf{u}_y + \sin(\omega t - kz)\mathbf{u}_x] \tag{2.61}$$

In order to examine the polarisation character of this new wave, let us study the curve described by the tip of the electric field vector at a fixed plane, for instance, the plane defined by $z = 0$. At this position, the time dependence of the fields is:

$$\mathcal{E}_x = E_0 \cos \omega t \quad \text{and} \quad \mathcal{E}_y = -E_0 \sin \omega t \tag{2.62}$$

$$\mathcal{H}_x = H_0 \sin \omega t \quad \text{and} \quad \mathcal{H}_y = H_0 \cos \omega t \tag{2.63}$$

The modulus of the electric field vector is therefore:

$$\mathcal{E}^2 = \mathcal{E}_x^2 + \mathcal{E}_y^2 = E_0^2 \tag{2.64}$$

and for the magnetic field:

$$\mathcal{H}^2 = \mathcal{H}_x^2 + \mathcal{H}_y^2 = H_0^2 \tag{2.65}$$

which indicates that, at a fixed plane, the tip of the electric field vector (and the magnetic field vector) describe a circle. For that reason, this wave is said to be *circularly polarised*. Moreover, looking at the wave along the propagation direction, one can observe that the electric field vector rotates contra-clockwise, and thus we are dealing with a *left-hand circularly polarised* wave.

On a general form, if two linearly polarised waves, mutually perpendicular, are superposed, having the same propagation direction and frequency, but with different amplitudes and relative phases, at a generic plane (for instance, at $z = 0$), we will have:

$$\mathcal{E}_x = E_{01} \cos(\omega t - \theta_1) \tag{2.66}$$

$$\mathcal{E}_y = -E_{02} \cos(\omega t - \theta_2) \tag{2.67}$$

For such a wave, the relation between the Cartesian components of the electric field is:

$$(\mathcal{E}_x/E_{01})^2 + (\mathcal{E}_y/E_{02})^2 - 2(\mathcal{E}_x/E_{01})(\mathcal{E}_y/E_{02}) \cos\theta = \sin^2\theta \tag{2.68}$$

where we have defined θ as the relative phase between $\mathcal{E}_x$ and $\mathcal{E}_y$ ($\theta \equiv \theta_2 - \theta_1$). This equation represents an ellipse, being the curve drawn by the electric field, and describes an *elliptically polarised wave*.

In general, the principal axis of the ellipse will be tilted with respect to the x and y axis. In particular, for $\theta = \pi/2, 3\pi/2, \ldots$, the major and minor axis of the ellipse will lie along the x and y axis. In this case, if in addition the amplitude of the components are equal ($E_{01} = E_{02}$), then the ellipse will degenerate into a circle. For relative phase of $\theta = 0, \pi, 2\pi, \ldots$, the ellipse will become a straight line, with:

$$\mathcal{E}_x = \pm(E_{02}/E_{01})\mathcal{E}_y \tag{2.69}$$

which represents once again a linearly polarised wave.

2.1.6 Light propagation in absorbing media

An absorbing medium is characterised by the fact that the energy of the EM radiation is dissipated in it. This would imply that the amplitude of a plane EM wave decreases exponentially as the wave propagates along the absorbing medium. The mathematical description of light propagation in absorbing media can be treated by considering that the dielectric permittivity is no longer a real number, but a complex quantity ε_c. In terms of field description, this implies that the electric displacement, now related by the electric field by $\mathcal{D} = \varepsilon_c \mathcal{E}$, will not be in phase with the electric field in general. As the refractive index was defined as a function of the dielectric permittivity, it will be in general a complex number itself, now defined by:

$$n_c = \sqrt{\frac{\varepsilon_c}{\varepsilon_0}} \tag{2.70}$$

where n_c is called the *complex refractive index*. It is useful to work with the real and imaginary part separately, and in this way we define:

$$n_c = n - i\kappa \tag{2.71}$$

where now n is the *real refractive index*, and κ is called the *absorption index*.

In addition, from the Helmholtz equation (2.44) the relation between the *complex wavevector* $\mathbf{k}_c$ (now complex) and the complex refractive index n_c is

$$\mathbf{k}_c^2 \equiv \omega^2 \varepsilon_c \mu = n_c^2 k_0 \tag{2.72}$$

Because the wavevector is now a complex vector, we can separate its real and imaginary parts in the following way:

$$\mathbf{k}_c \equiv \mathbf{k} - i\mathbf{a} \tag{2.73}$$

where $\mathbf{k}$ represents the real wavevector, and $\mathbf{a}$ is called the attenuation vector. The relation between the vectors $\mathbf{k}$ and $\mathbf{a}$ and the optical constant of the material medium n and κ are deduced from equation (2.72), resulting in

$$\mathbf{k}^2 - \mathbf{a}^2 = k_0^2(n^2 - \kappa^2) \tag{2.74}$$

$$\mathbf{ka} = k_0^2 n\kappa \tag{2.75}$$

Taking into account these definitions, the electric field for a plane monochromatic wave in absorbing medium will have the general form:

$$\mathcal{E}(\mathbf{r}, t) = Re\left[\mathbf{E}_0 e^{i(\omega t - \mathbf{k}_c \mathbf{r})}\right] = Re\left[\mathbf{E}_0 e^{-\mathbf{ar}} e^{i(\omega t - \mathbf{kr})}\right] \tag{2.76}$$

The planes of constant amplitude will be determined by the condition $\mathbf{ar}$ = constant, and therefore they will be planes perpendicular to the attenuation vector $\mathbf{a}$. On the other hand, the planes of equal phase will be defined by the condition of $\mathbf{kr}$ = constant, and thus the phase front will be planes perpendicular to the real wavevector $\mathbf{k}$. In general, these two planes will not be coincident, and in this case the EM wave is said to be an *inhomogeneous wave*.

Nevertheless, the most common situation faced in light propagation in absorbing media is the case where the vectors $\mathbf{k}$ and $\mathbf{a}$ are parallel, and such a wave is called a *homogeneous wave*. In this particular case, the vectors $\mathbf{k}_c$, $\mathbf{k}$ and $\mathbf{a}$ are related to the optical constant of the medium through the following simple relations:

$$\mathbf{k} = n\mathbf{k}_0 \tag{2.77}$$

$$\mathbf{a} = \kappa\mathbf{k}_0 \tag{2.78}$$

$$\mathbf{k}_c \equiv (n - i\kappa)\mathbf{k}_0 \tag{2.79}$$

and the electric field takes the form:

$$\mathcal{E}(\mathbf{r}, t) = Re\left[\mathbf{E}_0 e^{i(\omega t - n_c \mathbf{k}_0 \mathbf{r})}\right] = Re\left[\mathbf{E}_0 e^{-\kappa \mathbf{k}_0 \mathbf{r}} e^{i(\omega t - n\mathbf{k}_0 \mathbf{r})}\right] \tag{2.80}$$

This expression describes a wave propagating in the direction defined by the vector $\mathbf{k}_0$, with a phase velocity given by $v = c/n$, and whose amplitude decreases exponentially in the same direction to that propagation.

One important aspect concerning light propagation in absorbing media is the intensity variation suffered by the wave as it propagates. To calculate the intensity associated with the wave we use the expressions given in equations (2.38) and (2.39), as well as the result obtained in (2.80). Without loss of generality, we assume that the propagation is along the z-axis; in this case, the intensity takes the form:

$$I(z) = \frac{1}{2c\mu_0}|\mathbf{E}_0|^2 e^{-2\kappa k_0 z} \tag{2.81}$$

If we now define I_0 as the intensity associated with the wave at the plane $z = 0$, it follows:

$$I_0 = \frac{1}{2c\mu_0}|\mathbf{E}_0|^2 \tag{2.82}$$

and the expression for $I(z)$ becomes in a more compact form:

$$I(z) = I_0 e^{-2\kappa k_0 z} \tag{2.83}$$

This formula indicates that the intensity of the wave decreases exponentially as a function of the propagation distance.

In some applications it is convenient to deal with the absorption by using the *absorption coefficient* α, defined as:

$$\alpha \equiv 2\kappa k_0 = 2\kappa\omega/c \tag{2.84}$$

which has dimensions of m^{-1}. In this way, the attenuation of a light beam passing an absorbing medium is expressed in a compact manner by:

$$I(z) = I_0 e^{-\alpha z} \tag{2.85}$$

When working with optical fibres or optical waveguides, it is usual to refer to the light attenuation in *decibels* (dB), whose relation with the absorption coefficient is:

$$1 \text{ dB} \equiv 10 \log_{10}(I_0/I) = 4.3\ \alpha d \tag{2.86}$$

where I/I_0 represents the fraction of light intensity after a distance d.

In the case of metallic media, characterised by having a high electrical conductivity σ (compared with $\varepsilon\omega$), it is necessary to use the complete wave equations given in (2.20) and (2.21), that include the term corresponding to the first time derivative of the electric (and magnetic) field. Fortunately, when dealing with monochromatic waves, it is possible to include the electrical conductivity as an extra contribution to the dielectric permittivity in the following way:

$$\varepsilon_G = \varepsilon - i\sigma/\omega \tag{2.87}$$

where ε_G (clearly, a complex quantity) is known as generalised dielectric permittivity. By using this form, the Helmholtz equation is still valid, expressed as:

$$\nabla^2 U(\mathbf{r}) + k_c^2 U(\mathbf{r}) = 0 \tag{2.88}$$

defining $\mathbf{k}_c$ in this case as:

$$\mathbf{k}_c^2 = \omega^2 \varepsilon_G \mu = \omega^2(\varepsilon - i\sigma/\omega)\mu = n_c^2 k_0 \tag{2.89}$$

In this way, the mathematical formalism of the metallic media is equivalent to that of the absorbing media, by the only consideration of including the electric conductivity on the generalised dielectric permittivity through the equation (2.89).

Finally, it is important to note that, even in a totally transparent dielectric medium ($\kappa = 0$), it is possible not to have null values for the attenuation vector **a**; such is the situation found in the phenomenon of *total internal reflection*. This particular case, of capital importance in the description and performance of integrated photonics devices, will be dealt with in the next chapter in connection with light propagation in optical waveguides. From equation (2.75) it is evident that this situation only is possible if vectors **a** and **k** are perpendicular. As we have discussed earlier, in this case we are dealing with an inhomogeneous wave, where its planes of equal amplitude and equal phase are perpendicular.

2.2 EM Waves at Planar Dielectric Interfaces

2.2.1 Boundary conditions at the interface

Up to now we have described the propagation of EM waves in free space or through a material medium. Another important aspect in the study of light propagation is the behaviour of EM waves passing from one medium to another. We will analyse this by studying the behaviour of an EM monochromatic plane wave travelling through a homogeneous medium, incident on a second homogeneous medium, separated from the former by a planar interface. We will see that, besides the existence of a transmitted wave in the second medium, the incident wave partially reflects at the interface, giving rise to a reflected wave. The equations that determine the reflection and transmission coefficients can be studied separately in two groups: in one situation, the electric field of the incident EM wave has only a parallel component with respect to the incident plane (the magnetic field being perpendicular to that plane); the other group refers to incident EM waves in which the electric vector has only the component perpendicular to the incident plane, and therefore the magnetic vector is perpendicular to that plane. These two cases are mutually independent, and can be treated separately: from them it is possible to deduce the equations that govern reflection and transmission for any plane wave with arbitrary polarisation state.

The relations between the incident, reflected and transmitted waves are obtained by setting the adequate boundary conditions for the fields at the planar interface, which are derived directly from Maxwell's equations. Because the $\mathcal{E}$, $\mathcal{D}$, $\mathcal{H}$ and $\mathcal{B}$ fields are not independent, but related by Maxwell's equations and the constitutive relations of the media, only some of the boundary conditions should be taken into account.

From equations (2.9) and (2.10), one obtains respectively that the normal components of the fields $\mathcal{D}$ and $\mathcal{B}$ should be kept across the boundary, that is:

$$(\mathcal{D}^{\text{Normal}})_{\text{Medium 1}} = (\mathcal{D}^{\text{Normal}})_{\text{Medium 2}} \quad \text{at the interface} \tag{2.90}$$

$$(\mathcal{B}^{\text{Normal}})_{\text{Medium 1}} = (\mathcal{B}^{\text{Normal}})_{\text{Medium 2}} \quad \text{at the interface} \tag{2.91}$$

On the other hand, by using Maxwell's equations (2.11) and (2.12) respectively, the conditions of continuity across the interface of the tangential components of the $\mathcal{E}$ and $\mathcal{H}$ fields are obtained:

$$(\mathcal{E}^{\text{Tangential}})_{\text{Medium 1}} = (\mathcal{E}^{\text{Tangential}})_{\text{Medium 2}} \quad \text{at the interface} \tag{2.92}$$

$$(\mathcal{H}^{\text{Tangential}})_{\text{Medium 1}} = (\mathcal{H}^{\text{Tangential}})_{\text{Medium 2}} \quad \text{at the interface} \tag{2.93}$$

Let us consider an EM monochromatic plane wave, characterised by its angular frequency ω_{i} and wavevector $\mathbf{k}_{\mathrm{i}}$, incident from a homogeneous medium (1) to a planar frontier separating a different homogeneous medium (2). The dielectric media are characterised by their optical constant (ε_1, μ_1) and (ε_2, μ_2), where the subscript denotes the medium (1 or 2). If the two media are isotropic and homogeneous, the electric field vectors, using complex notation, corresponding to the incident, reflected and transmitted (or refracted) waves are expressed as:

$$\mathcal{E}_i(\mathbf{r}, t) = \mathbf{E}_i e^{i(\omega_i t - \mathbf{k}_i \mathbf{r})} \tag{2.94}$$

$$\mathcal{E}_r(\mathbf{r}, t) = \mathbf{E}_r e^{i(\omega_r t - \mathbf{k}_r \mathbf{r})} \tag{2.95}$$

$$\mathcal{E}_t(\mathbf{r}, t) = \mathbf{E}_t e^{i(\omega_t t - \mathbf{k}_t \mathbf{r})} \tag{2.96}$$

where $\mathbf{k}_{\mathrm{r}}$ and $\mathbf{k}_{\mathrm{t}}$ are the wavevectors of the reflected and transmitted waves, and ω_{r} and ω_{t} are their respective angular frequencies. The vectors $\mathbf{E}_i$, $\mathbf{E}_r$ and $\mathbf{E}_t$ represent the electric field complex amplitudes of the incident, reflected and transmitted waves, respectively, being independent of the time and space. The magnetic fields vectors associated with each wave have similar expressions, and can be deduced from equation (2.51).

Let us apply the condition of the continuity of the tangential component of the electric field across the interface. The condition (2.92) as well as the expressions for the electric fields gives the following relation:

$$[\mathcal{E}_i(\mathbf{r}, t) + \mathcal{E}_r(\mathbf{r}, t)]^{\text{Tangential}} = [\mathcal{E}_t(\mathbf{r}, t)]^{\text{Tangential}} \tag{2.97}$$

or more explicitly, using the fact that the waves are monochromatic plane waves:

$$[\mathbf{E}_i e^{i(\omega_i t - \mathbf{k}_i \mathbf{r})} + \mathbf{E}_r e^{i(\omega_r t - \mathbf{k}_r \mathbf{r})}]^{\text{Tangential}} = [\mathbf{E}_t e^{i(\omega_t t - \mathbf{k}_t \mathbf{r})}]^{\text{Tangential}} \tag{2.98}$$

As this relation should be valid for any instant of time, it follows that:

$$\omega_{\mathrm{i}} = \omega_{\mathrm{r}} = \omega_{\mathrm{t}} \tag{2.99}$$

that indicates that the frequencies of the reflected and transmitted waves are equal to the frequency of the incident wave.

On the other hand, equation (2.98) should be valid for any point at the interface. If we choose the coordinate origin O at the interface, and the x axis in the normal direction to the plane boundary, then the position vector $\mathbf{r}$ would lie in this plane. Thus, the condition of equal spatial dependence on the exponents in equation (2.98) at the interface is expressed as:

$$k_{\mathrm{iy}}\, y + k_{\mathrm{iz}}\, z = k_{\mathrm{ry}}\, y + k_{\mathrm{rz}}\, z = k_{\mathrm{ty}}\, y + k_{\mathrm{tz}}\, z \quad \text{(at the interface } x = 0\text{)} \tag{2.100}$$

This result indicates that the tangential component of the wavevectors (for the incident, reflected and transmitted waves) must be equal:

$$[\mathbf{k}_{\mathrm{i}}]^{\mathrm{T}} = [\mathbf{k}_{\mathrm{r}}]^{\mathrm{T}} = [\mathbf{k}_{\mathrm{t}}]^{\mathrm{T}} \tag{2.101}$$

In other words, at the boundary only the perpendicular component of the wavevectors can change. Thus, the vectors $\mathbf{k}_{\mathrm{r}}$ and $\mathbf{k}_{\mathrm{t}}$ must lie in the plane defined by the $\mathbf{k}_{\mathrm{i}}$ vector

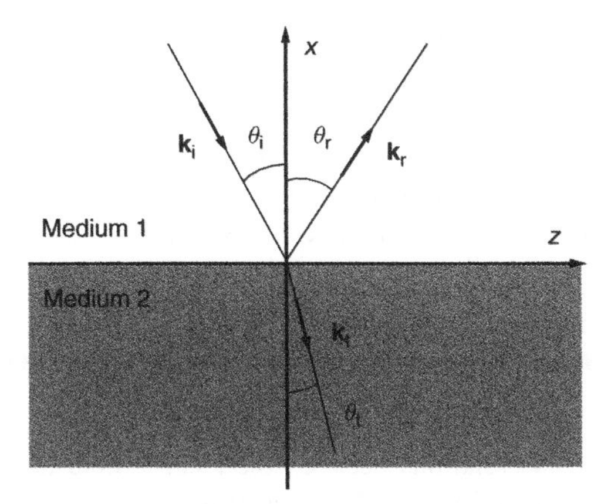

Figure 2.3 Reference system used to study the reflection and transmission experimented by a monochromatic plane wave incident from medium (1) to medium (2)

and the normal to the plane of the interface. This plane, perpendicular to the plane that separates both media, is called the *incident plane*, and all the wavevectors lie on it. These formulae are simplified if we choose the incident plane as the x-z plane, as indicated in Figure 2.3: in this case the y components of the wavevectors are null, and the relation (2.100) leads to:

$$k_{iz} = k_{rz} = k_{tz} \tag{2.102}$$

If now we introduce the incident angle θ_i, the reflected angle θ_r, and the transmitted angle θ_t, following Figure 2.3, the above equation can be expressed as:

$$k_i \sin\theta_i = k_r \sin\theta_r = k_t \sin\theta_t \tag{2.103}$$

As the electric fields given by (2.94)–(2.96) must be solutions of the wave equation, the wavevectors modulus should be:

$$k_i = \omega(\varepsilon_1\mu_1)^{1/2} = k_r \tag{2.104}$$

$$k_t = \omega(\varepsilon_2\mu_2)^{1/2} \tag{2.105}$$

From equations (2.103) and (2.104) one directly obtains:

$$\theta_i = \theta_r \tag{2.106}$$

that indicates that the reflected angle is equal to the incident angle, known as the *law of reflection*. Equation (2.103) also relates the transmitted and the incident waves, resulting in:

$$k_i \sin\theta_i = k_t \sin\theta_t \tag{2.107}$$

that is the mathematical expression for the *transmission law*. If the two homogeneous media are non-magnetic ($\mu_1 \approx \mu_2 \approx \mu_0$) and non-absorbing materials (real refractive indices), then:

$$(\varepsilon_1/\varepsilon_0)^{1/2} = (\varepsilon_{r1})^{1/2} = n_1 \tag{2.108}$$

$$(\varepsilon_2/\varepsilon_0)^{1/2} = (\varepsilon_{r2})^{1/2} = n_2 \tag{2.109}$$

and in this case the equation (2.107) takes the familiar form of:

$$n_1 \sin\theta_i = n_2 \sin\theta_t \tag{2.110}$$

which is the well-known *Snell's law*, valid for dielectric materials. In the case of absorbing media, the equation (2.102) is still valid, and is the correct relation to obtain the transmitted wave. In the most general case, this would give rise to an inhomogeneous transmitted wave.

2.2.2 Reflection and transmission coefficients: reflectance and transmittance

Now we will focus on the relations between the electric field amplitude for the incident, reflected and transmitted waves. In order to do this, we will use the appropriate boundary conditions (2.90)–(2.93) that should be fulfilled by the fields at the interface. We will consider two basic types of linearly polarised incident waves separately: the first deals with an EM wave in which the associated electric field vector lies on the incident plane; in the second the electric field vector is perpendicular to that plane. In the general case of an incident wave with an arbitrary polarisation state, the procedure is to decompose it into the two basic polarisations, treating them separately, and finally to re-compose the electric field by adding the two mutually orthogonal components. Using appropriate boundary conditions at the interface, it can be demonstrated that if the wave has its electric field parallel to the incident plane, the reflected and transmitted waves will also have their electric field in that plane. In the same way, if the electric wave associated to the incident wave is perpendicular to the incident plane, the electric fields of the reflected and transmitted waves will also be perpendicular to the incident plane.

Let us consider the first case in which the electric field vector associated with the incident monochromatic plane wave lies on the incident plane, as depicted in Figure 2.4. As the wavevector is also on this plane, and the magnetic field vector is perpendicular to both vectors, it is deduced that the magnetic field vector must be perpendicular to the incident plane: this is the reason why this case is called *transverse magnetic incidence* (*TM incidence*). In this situation, the electric and magnetic fields are given by:

$$\mathcal{E}_i \equiv \mathcal{E}_i^{\parallel} \equiv [\mathcal{E}_{ix}, 0, \mathcal{E}_{iz}] \tag{2.111}$$

$$\mathcal{H}_i \equiv \mathcal{H}_i^{\perp} \equiv [0, \mathcal{H}_{iy}, 0] \tag{2.112}$$

where the symbols $\parallel$ and $\perp$ denote vectors parallel and perpendicular to the incident plane, respectively. As the electric field vector is parallel to the incidence plane, the TM incidence is also called *parallel incidence*. Applying the condition of the continuity of the tangential component of the electric field at the interface given by (2.92) we obtain:

$$\mathcal{E}_{iz} + \mathcal{E}_{rz} = \mathcal{E}_{tz} \tag{2.113}$$

and in terms of the incident, reflected and transmitted angles, using the geometry shown in Figure 2.4, we obtain:

$$[\mathrm{E}_i e^{i(\omega_i t - \mathbf{k}_i \mathbf{r})} \cos\theta_i - \mathrm{E}_r e^{i(\omega_r t - \mathbf{k}_r \mathbf{r})} \cos\theta_r]_{x=0} = [\mathrm{E}_t e^{i(\omega_t t - \mathbf{k}_t \mathbf{r})} \cos\theta_t]_{x=0} \tag{2.114}$$

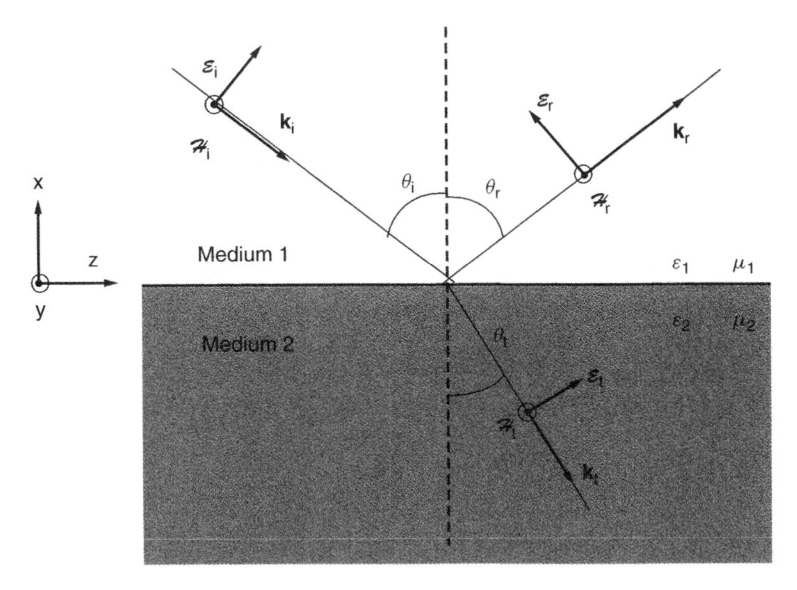

Figure 2.4 Reflection and transmission suffered by a plane wave incident on a planar interface that separates two dielectric media, showing the directions of the wavevector, electric field and magnetic field vectors for the incident wave (i), reflected wave (r) and transmitted wave (t). We have assumed that the electric field vector lies in the incidence plane (x–z plane), which corresponds to TM incidence or parallel incidence

where the three expressions should be evaluated at the interface ($x = 0$). As we have seen before, the temporal and spatial dependences of the exponentials are equal (at $x = 0$), and therefore it follows that:

$$E_\mathrm{i} \cos\theta_\mathrm{i} - E_\mathrm{r} \cos\theta_\mathrm{i} = E_\mathrm{t} \cos\theta_\mathrm{t} \tag{2.115}$$

On the other hand, the condition of continuity of the normal component of the dielectric displacement vector (2.90) is expressed in this geometry as:

$$\mathcal{D}_\mathrm{ix} + \mathcal{D}_\mathrm{rx} = \mathcal{D}_\mathrm{tx} \tag{2.116}$$

and taking into account the constitutive relation (2.13), we can express this relation as a function of the electric fields:

$$\varepsilon_1 E_\mathrm{i} \sin\theta_\mathrm{i} + \varepsilon_1 E_\mathrm{r} \sin\theta_\mathrm{i} = \varepsilon_2 E_\mathrm{t} \sin\theta_\mathrm{t} \tag{2.117}$$

By combining the conditions (2.115) and (2.117), we obtain the following expression for the relation between the electric field amplitudes of the reflected and incident waves:

$$r_\mathrm{TM} \equiv \frac{E_r}{E_i} = \frac{n_2 \cos\theta_i - n_1 \cos\theta_t}{n_2 \cos\theta_i + n_1 \cos\theta_t} \tag{2.118}$$

where r_TM denotes the *reflection coefficient* for parallel polarisation. In a similar way, the relation between the amplitude between the transmitted and incident waves is obtained:

$$t_\mathrm{TM} \equiv \frac{E_t}{E_i} = \frac{2n_1 \cos\theta_i}{n_1 \cos\theta_t + n_2 \cos\theta_i} \tag{2.119}$$

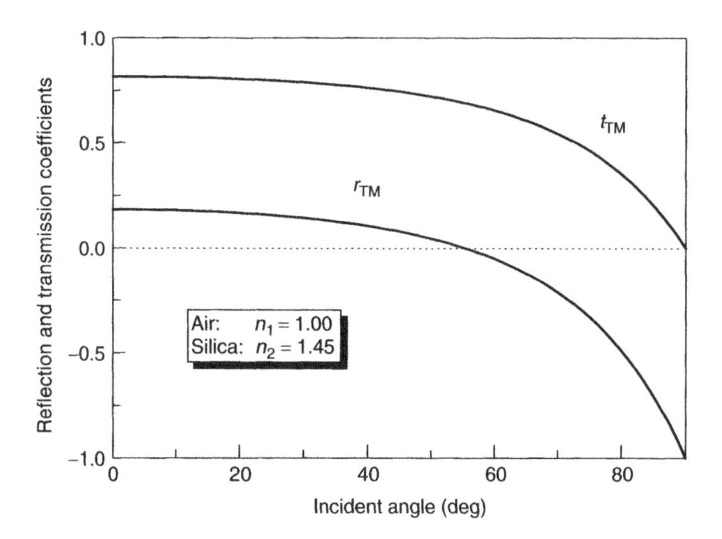

Figure 2.5 Reflection and transmission coefficients for TM-incidence (or parallel polarisation) in the case of air–silica ($n_1 = 1.00$, $n_2 = 1.45$)

being here t_{TM} the *transmission coefficient* for parallel polarisation. Figure 2.5 shows the r_{TM} and t_{TM} values as function of the incident angle in the case of air–silica ($n_1 = 1.00, n_2 = 1.45$). It is convenient to recall that in this particular case the r_{TM} and t_{TM} coefficients are real, but in general r_{TM} and t_{TM} can be complex magnitudes.

Although the reflection and transmission coefficients give us valuable information concerning the relation between the electric field amplitudes of the incident, reflected and transmitted waves, in many cases the relevant parameter is the fraction of the incident energy that is reflected and transmitted at the interface, defined through reflectance and transmittance. In particular, the *reflectance* R is defined as the quotient between the reflected energy in an unit of time over a differential area, and the incident energy per unit of time over the same area at the interface. Similarly, the *transmittance* is defined as the quotient between the transmitted energy per unit of time over a differential area and the incident energy in that unit of time over the same area. These fluxes of energy are related to the Poynting vector component perpendicular to the plane of the interface, or according to the geometry in Figure 2.3, the flux along the x-axis. The expressions for the reflectance and transmittance are therefore expressed as:

$$R = \frac{|\langle \mathbf{S}_{ref} \cdot d\mathbf{a} \rangle|}{|\langle \mathbf{S}_{inc} \cdot d\mathbf{a} \rangle|} = \frac{|\langle (\mathbf{S}_{ref})_x \rangle|}{|\langle (\mathbf{S}_{inc})_x \rangle|} \tag{2.120}$$

$$T = \frac{|\langle \mathbf{S}_{trans} \cdot d\mathbf{a} \rangle|}{|\langle \mathbf{S}_{inc} \cdot d\mathbf{a} \rangle|} = \frac{|\langle (\mathbf{S}_{trans})_x \rangle|}{|\langle (\mathbf{S}_{inc})_x \rangle|} \tag{2.121}$$

where $\mathbf{S}_{inc}$, $\mathbf{S}_{ref}$ and $\mathbf{S}_{trans}$ represent the Poynting vectors associated with the incident, reflected and transmitted waves, and $d\mathbf{a}$ represents the differential area at the interface.

Taking into account the definition of the Poynting vector given in (2.30), and the time average of the cosine squared function, we obtain:

$$|\langle S_x \rangle| = |\langle (\mathcal{E} x \mathcal{H})_x \rangle| = \tfrac{1}{2}(k/\omega\mu)|E|^2 \cos\theta \tag{2.122}$$

where θ is the angle formed between the vectorial product $\mathcal{E} \times \mathcal{H}$ (that is, the wavevector $\mathbf{k}$) with the x-axis. Using this expression and combining it with (2.120), we obtain the reflectance for TM incidence:

$$R_{\text{TM}} = \frac{(1/2)(\varepsilon_1/\mu_1)^{1/2}|E_r|^2 \cos\theta_r}{(1/2)(\varepsilon_1/\mu_1)^{1/2}|E_i|^2 \cos\theta_i} = \left|\frac{E_r}{E_i}\right|^2 = |r_{\text{TM}}|^2 \tag{2.123}$$

or in terms of the incident and transmitted angles:

$$R_{\text{TM}} = \left(\frac{n_2 \cos\theta_i - n_1 \cos\theta_t}{n_2 \cos\theta_i + n_1 \cos\theta_t}\right)^2 \tag{2.124}$$

In a similar way, the transmittance is expressed as function of the refractive indices and the incident and transmitted angles as:

$$T_{\text{TM}} = \frac{4n_1 n_2 \cos\theta_i \cos\theta_t}{(n_2 \cos\theta_i + n_1 \cos\theta_t)^2} \tag{2.125}$$

In addition, from (2.124) and (2.125) one can readily obtain:

$$R_{\text{TM}} + T_{\text{TM}} = 1 \tag{2.126}$$

that can be considered as the electromagnetic energy conservation for the incident, reflected and transmitted waves at the interface.

From equation (2.124) it follows that the reflectance will vanish for the condition $n_2 \cos\theta_{\text{i}} = n_1 \cos\theta_{\text{t}}$. Apart from the obvious case in which the two media are optically equivalent ($n_1 = n_2$), there exists another particularly interesting situation in which $R_{\text{TM}} = 0$. By combining (2.124) with the Snell's law, one obtains that the reflectance is zero for an incident angle that fulfils the equation:

$$\text{tg}\theta_{\text{i}} = n_2/n_1 \tag{2.127}$$

This angle, for which $R_{\text{TM}} = 0$, is called *Brewster's angle* θ_{B} or the *polarising angle*, because the reflected wave will be linearly polarised for an incident wave with arbitrary polarisation state. Figure 2.6 shows the reflectance and transmittance curves for the air–silica case, where we observe that Brewster's angle is situated at $\theta_{\text{B}} = 55.4^\circ$.

For the particular case of normal incidence ($\theta_{\text{i}} = 0$), the formula for the reflectance is simplified to:

$$R = \left(\frac{n_2 - n_1}{n_2 + n_1}\right)^2 \tag{2.128}$$

where now we have omitted the subscript TM, because for normal incidence there is no physical difference between parallel or perpendicular incidence.

For the case air–glass the reflectance at normal incidence is $R \approx 5\%$, which is a relatively low value. Nevertheless, for materials with higher refractive indices (f.i. $LiNbO_3$, $n \approx 2.2$; GaAs, $n \approx 3.4$, see Table 2.1), the reflectance at normal incidence are high enough ($R(LiNbO_3) \approx 14\%$, $R(\text{GaAs}) \approx 30\%$) to be used as partially reflecting mirrors in some integrated photonic components, such as in propagating loss measurement by the Fabry-Perot method [1], or to provide feedback in integrated waveguide lasers [2].

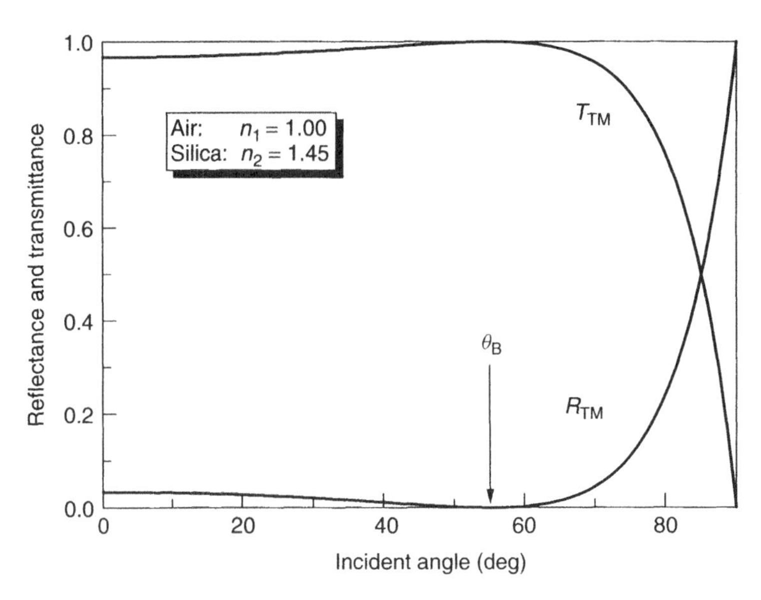

Figure 2.6 Reflectance and transmittance for TM incidence corresponding to the interface air–silica ($n_1 = 1.00$, $n_2 = 1.45$). For an incident angle at $\theta_i = \theta_B$ the reflectance vanishes, corresponding to an angle of 55.4°

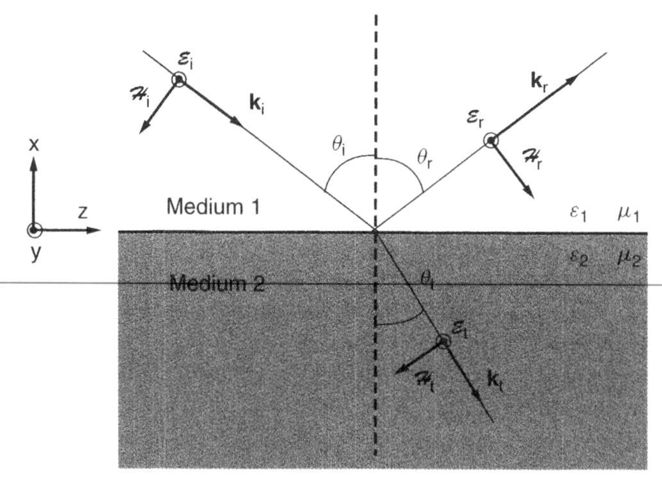

Figure 2.7 Reflection and transmission corresponding to TE incidence (perpendicular polarisation). While the electric field vectors are perpendicular to the incident plane (x–z plane), the wavevectors and the magnetic field vectors lie in that plane

We now consider the situation in which the electric field vector of the incident wave is perpendicular to the incident plane; therefore, this case is called *transverse electric incidence* (*TE incidence*) or *perpendicular polarisation*. Figure 2.7 depicts this case, where the electric field vectors are perpendicular to the incident plane, and the magnetic field vectors lie in that plane. Taking into account the reference system of the figure, the electric and magnetic field vectors associated with the incident wave are:

$$\mathcal{E}_i \equiv \mathcal{E}_i^{\perp} \equiv [0, \mathcal{E}_{iy}, 0] \tag{2.129}$$

$$\mathcal{H}_i \equiv \mathcal{H}_i^{\parallel} \equiv [\mathcal{H}_{ix}, 0, \mathcal{H}_{iz}] \tag{2.130}$$

The continuity of the tangential component of the electric field across the boundary (2.92) is expressed in this case as:

$$E_{iy} + E_{ry} = E_{ty} \tag{2.131}$$

To obtain the reflection and transmission coefficients it is necessary to find a second relation between the electric field amplitudes. This is obtained by imposing the condition of continuity of the tangential component of the magnetic field vector (2.93) at the interface:

$$H_{iz} + H_{rz} = H_{tz} \tag{2.132}$$

and by relating the magnetic field vectors with the electric field vectors by using equation (2.51). After straightforward calculations, the boundary condition (2.132) becomes:

$$k_{ix}(E_{iy} - E_{ry}) = k_{tx} E_{ty} \tag{2.133}$$

By combining (2.131) with (2.133), the reflection and transmission coefficients for TE incidence are obtained as a function of the wavevectors:

$$r_{TE} \equiv \frac{E_r}{E_i} = \frac{k_{ix} - k_{tx}}{k_{ix} + k_{tx}} \tag{2.134}$$

$$t_{TE} \equiv \frac{E_t}{E_i} = \frac{2k_{ix}}{k_{ix} + k_{tx}} \tag{2.135}$$

These coefficients can be expressed in a more convenient form as a function of the incident and refracted angles and the refractive indices of the two media by using Snell's law:

$$r_{TE} = \frac{n_1 \cos\theta_i - n_2 \cos\theta_t}{n_1 \cos\theta_i + n_2 \cos\theta_t} \tag{2.136}$$

$$t_{TE} = \frac{2n_1 \cos\theta_i}{n_1 \cos\theta_i + n_2 \cos\theta_t} \tag{2.137}$$

Figure 2.8 shows the reflection and transmission coefficients as a function of the incident angle in the case of air–silica interface for TE incidence, where both coefficients are real in the whole range of incident angles. As can be seen, the transmission coefficient is positive, indicating that the direction of the electric field vector of the transmitted wave is coincident to that of the incident wave. By contrast, the electric field vector associated with the reflected wave is reversed in respect to that of the incident wave, indicating a phase shift of π in the reflected wave.

Following a similar form used for TM incidence, the reflectance and the transmittance for TE incidence are expressed as:

$$R_{TE} = \left(\frac{n_1 \cos\theta_i - n_2 \cos\theta_t}{n_1 \cos\theta_i + n_2 \cos\theta_t}\right)^2 \tag{2.138}$$

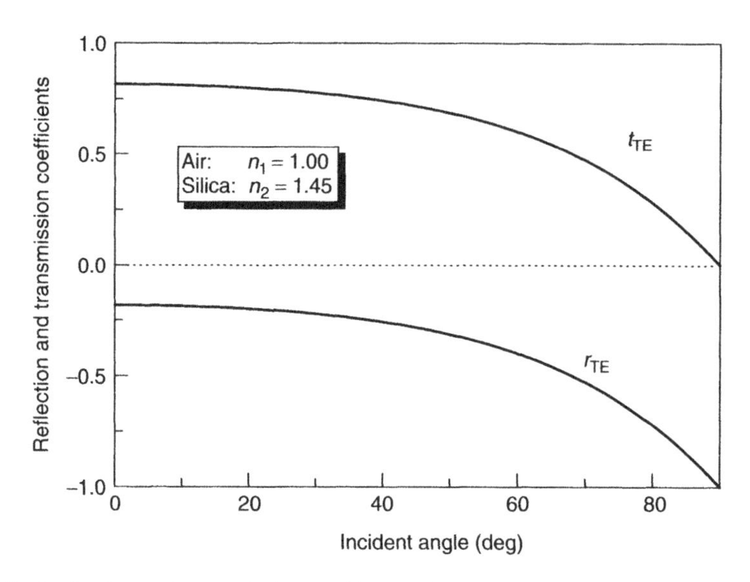

Figure 2.8 Reflection and transmission coefficients for TE incidence at the air–silica boundary

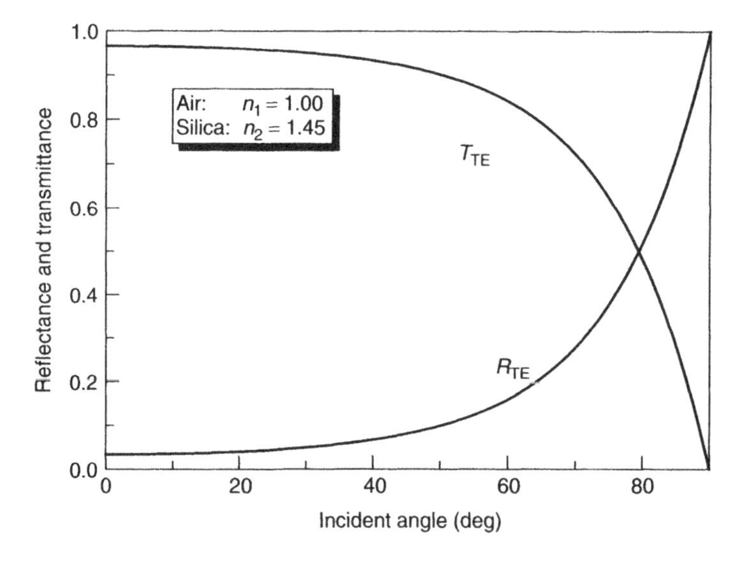

Figure 2.9 Reflectance and transmittance for TE incidence in the air–silica interface

$$T_{TE} = \frac{4n_1 \cos\theta_i n_2 \cos\theta_t}{(n_1 \cos\theta_i + n_2 \cos\theta_t)^2} \tag{2.139}$$

The reflectance and transmittance of a TE-polarised wave incident from air to silica are plotted in Figure 2.9. At variance to that found in TM incidence, in TE incidence the reflectance is a monotonous increasing function of the incident angle. Therefore, if a beam of non-polarised light is incident at an angle of θ_B, the interface only will reflect the TE component of such radiation, and thus the reflected wave will be linearly polarised with the electric field vector perpendicular to the incident plane. This is the reason why Brewster's angle is also called the polarising angle, and this phenomenon can be used to design polarisation devices.

2.2.3 Total internal reflection

The formulae describing the reflectance and transmittance whether for TM incidence or TE incidence, were obtained by assuming that the light is incident from a less dense medium (1) to a denser medium (2), or that the refractive index of medium (1) is lower that the refractive index of medium (2) ($n_1 < n_2$). This is called *soft incidence* (formulae (2.110), (2.137), (2.138)). From Snell's law (2.110), if $n_1 < n_2$ holds, it is easy to show that, regardless of the incident angle θ_i, the refracted angle always will exist, or in other words, the refracted angle θ_t, will always be a real number.

In contrast, if the plane wave is incident from a denser to a less dense medium ($n_1 > n_2$, *hard incidence*) an exceptional phenomenon takes place for a certain range of incident angles, for which the formulae formerly given for R and T can no longer be applied. For hard incidence ($n_1 > n_2$) there exists an incident angle θ_i for which the refracted angle θ_t takes the value of $\pi/2$ radians. This angle is called the *critical angle* θ_c, and its value, calculated directly from Snell's law is:

$$\theta_c \equiv \sin^{-1}(n_2/n_1) \tag{2.140}$$

For incident angles higher than the critical angle, the sine of the refracted angle will reach values greater than 1, thus the refracted angle is no longer a real number according to Snell's law. Nevertheless, this does not imply that in medium (2) there is no transmitted wave, as we will show.

In order to calculate the reflectivity in a case of hard incidence, it is necessary to evaluate $\cos\theta_t$ included in the formulae for the reflection and transmission coefficients. According to Snell's law, it follows that:

$$\cos\theta_t = -(1 - \sin^2\theta_t)^{1/2} \tag{2.141}$$

where the negative sign of the squared root has been chosen so that the complete expression for the electric field of the transmitted wave has a correct physical meaning. Taking into account that now $\sin\theta_t > 1$, the last formula can be expressed as:

$$\cos\theta_t = -\mathrm{i}(\sin^2\theta_t - 1)^{1/2} = -\mathrm{i}B \tag{2.142}$$

where the magnitude B has been defined as a real number by:

$$B \equiv (\sin^2\theta_t - 1)^{1/2} = (n_1^2\sin^2\theta_i/n_2^2 - 1)^{1/2} \tag{2.143}$$

According to the definition of B, the reflection and transmission coefficients for TM and TE incidence become:

$$r_{TM} = \frac{n_2\cos\theta_i + in_1B}{n_2\cos\theta_i - in_1B} = \frac{z_{TM}}{z_{TM}^*} = e^{i\phi_{TM}} \tag{2.144}$$

$$r_{TE} = \frac{n_1\cos\theta_i + in_2B}{n_1\cos\theta_i - in_2B} = \frac{z_{TE}}{z_{TE}^*} = e^{i\phi_{TE}} \tag{2.145}$$

Here, we have expressed the reflection coefficients as the quotient between a complex number z and its conjugated z^*, and thus the modulae of both reflection coefficients are 1. For that reason, the values of the reflectance for TM and TE polarisation are equal to 1:

$$|r_{TM}| = 1, \qquad R_{TM} = 1 \tag{2.146}$$

$$|r_{TE}| = 1, \qquad R_{TE} = 1 \tag{2.147}$$

where we have used the equation (2.123), and the fact that for perpendicular polarisation also holds that $R_{TE} = |r_{TE}|^2$.

On the other hand, the phase shift of the reflected waves, ϕ_{TM} and ϕ_{TE} in (2.144) and (2.145), are calculated from the following expressions:

$$\tan(\phi_{TM}/2) = n_1 B/n_2 \cos\theta_i \tag{2.148}$$

$$\tan(\phi_{TE}/2) = n_2 B/n_1 \cos\theta_i \tag{2.149}$$

The magnitudes of the phase shifts ϕ_{TE} and ϕ_{TM} are very important when establishing the condition of light propagation in planar optical waveguides, which gives rise to the calculation of the allowed propagating modes, as we will see in the next chapter. These phase shifts are represented in Figure 2.10, for the case of total internal reflection in the boundary silica–air ($n_1 = 1.45$, $n_2 = 1.00$). Note that for angles lower than the critical angle, the phase shifts are either 0 or π, but for incident angles greater that the critical angle, the phase shift is a monotonous increasing function of the angle.

In order to obtain the complete expression for the wave in the second medium, it is convenient to write explicitly the x and z components of the wavevector for the transmitted wave, given by:

$$k_{tx} = -k_t \cos\theta_t = -k_0 n_2 \cos\theta_t = ik_0 n_2 B \tag{2.150}$$

$$k_{tz} = k_{iz} = k_0 n_1 \sin\theta_i \tag{2.151}$$

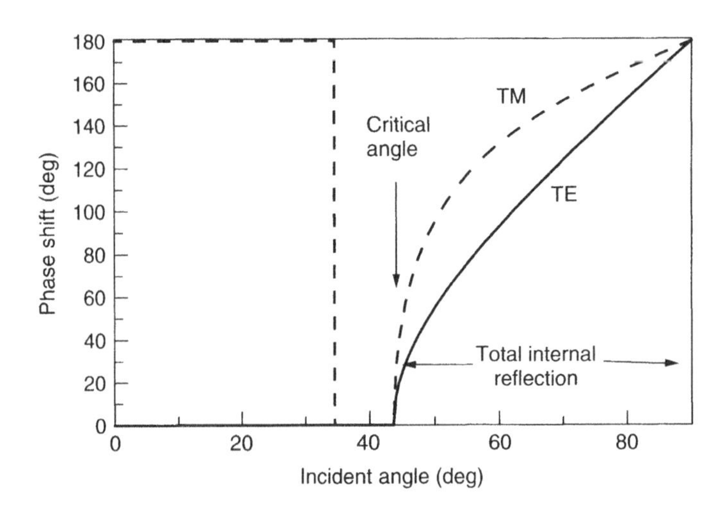

Figure 2.10 Phase shift experienced by the reflected waves for TM and TE incidence produced by effect of total internal reflection, in the case of the silica–air interface

By using the definition of the complex wavevector given in (2.73), the real wavevector $\mathbf{k}_t$ and the attenuation vector $\mathbf{a}_t$ of the transmitted wave are expressed by:

$$\mathbf{k}_t = [0, 0, k_0 n_1 \sin\theta_i] \tag{2.152}$$

$$\mathbf{a}_t = [k_0 n_2 B, 0, 0] \tag{2.153}$$

showing that the vectors $\mathbf{k}_t$ and $\mathbf{a}_t$ are mutually perpendicular, and as we saw in section (2.1.6), this situation corresponds to the case of an inhomogeneous wave. Taking into account the expression (2.80), the electric field associated with the transmitted wave is of the form:

$$\mathbf{E}_t(\mathbf{r}, t) = Re\{\mathbf{E}_0 e^{-\mathbf{a}_t\mathbf{r}} e^{i(\omega t - \mathbf{k}_t\mathbf{r})}\} = Re\{\mathbf{E}_0 e^{k_0 n_2 B x} e^{i(\omega t - k_0 n_1 \sin\theta_i z)}\} \tag{2.154}$$

This expression indicates that in the second medium there exists an EM wave that propagates along the interface (z-axis), and decreases in the perpendicular direction to the interface (x-direction, $x < 0$). The propagation along the z-direction does not depend on the medium (2), because it is controlled by $k_0 n_1 \sin\theta_i$. The amplitude of the electric field in the medium (2) decreases exponentially with the distance x in respect to the boundary. precisely for this reason, the fact that the wave in medium (2) shows an exponential decay with the distance, this wave is known as an *evanescent wave*. The *penetration depth*, defined as the distance to the interface where the electric field decreases a factor e, is given by:

$$x_{1/e} = 1/a_t = 1/(k_0 n_2 B) \approx 1/k_0 n_2 = \lambda_0/2\pi n_2 \tag{2.155}$$

where the last result has been obtained by assuming that $B \sim 1$, and indicates that the electric field penetrates in the medium (2) to approximately a distance of a tenth of the wavelength.

Although in the medium (2) there exists a transmitted wave given by the expression (2.154), it can be demonstrated that there is no energy flux in the x-direction [3], indicating that the light is totally reflected at the boundary, although a small fraction of it penetrates in the medium (2):

$$\langle S_x \rangle = \langle \mathcal{E}_y \mathcal{H}_z \rangle = 0 \tag{2.156}$$

By contrast, there is energy flow along the z-direction, parallel to the interface that separates both media, as it is shown in Figure 2.11. In the case of TE incidence, the z-component of the energy flux in the medium (2) is expressed as:

$$\langle S_z \rangle = \langle -\mathcal{E}_y \mathcal{H}_x \rangle = \sqrt{\frac{\mu_0}{\varepsilon_0}} \frac{n_1 \sin\theta_i}{2n_2^2} |t_{TE}|^2 E_1^2 e^{2k_0 n_2 B x} \tag{2.157}$$

where the squared modulus of t_{TE} should be evaluated by using the equations (2.137), (2.142) and (2.143). This result is of special importance when examining the behaviour of light confined in waveguides, which is the subject of the next chapter. We will show that, although a waveguide confines the light by total internal reflection in its adjacent interfaces, EM radiation would be found in the surrounding media. This fact can be exploited favourably to couple light in waveguides [4], to fabricate waveguide sensors [5], etc., and is one of the key factors in the modal coupling as we will see in the next chapters. Table 2.2 shows the magnitudes introduced in this chapter.

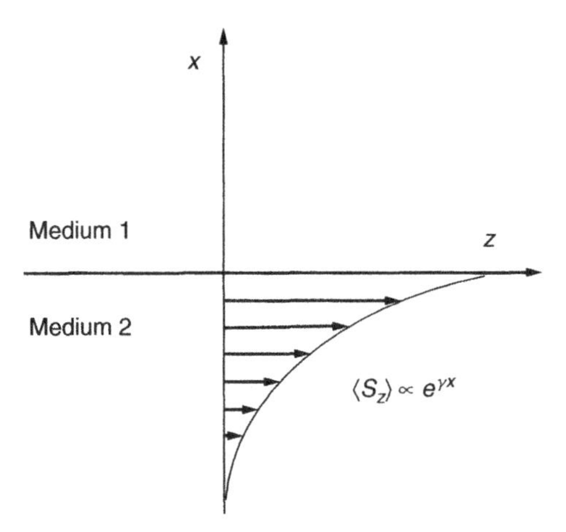

Figure 2.11 Penetration of the evanescent wave in the medium (2), when a plane wave is incident from the medium (1) with an incident angle greater than the critical angle

Table 2.2 Magnitudes introduced in Chapter 2, including the symbols used and their units in the SI

Magnitude	Symbol	Unit (S.I.)
Electric field	E	$\mathrm{V\,m^{-1}}$, $\mathrm{N\,C^{-1}}$, $\mathrm{m\,Kg\,s^{-2}\,C^{-1}}$
Magnetic field	H	$\mathrm{A\,m^{-1}}$, $\mathrm{m^{-1}\,s^{-1}\,C}$
Electric displacement	D	$\mathrm{m^{-2}\,C}$
Magnetic induction	B	$\mathrm{Wb\,m^{-2}}$, $\mathrm{Kg\,s^{-1}\,C^{-1}}$
Current density	J	$\mathrm{A\,m^{-2}}$, $\mathrm{m^{-2}\,s^{-1}\,C}$
Charge density	ρ	$\mathrm{m^{-3}\,C}$
Poynting vector	S	$\mathrm{W\,m^{-2}}$, $\mathrm{m^{2}\,Kg\,s^{-3}}$
Intensity	I	$\mathrm{W\,m^{-2}}$, $\mathrm{m^{2}\,Kg\,s^{-3}}$
Angular frequency	ω	$\mathrm{rad\,s^{-1}}$
Frequency	ν	$\mathrm{s^{-1}}$, Hz
Period	T	s
Phase	φ	rad
Wavelength	λ	m
Permittivity	ε	$\mathrm{F\,m^{-1}}$, $\mathrm{m^{-3}\,Kg^{-1}\,s^{2}\,C^{2}}$
Permeability	μ	$\mathrm{H\,m^{-1}}$, $\mathrm{m\,Kg\,C^{-2}}$
Conductivity	σ	$\Omega^{-1}\mathrm{m^{-1}}$, $\mathrm{m^{-3}\,Kg^{-1}\,s\,C^{2}}$
Velocity, speed	v	$\mathrm{m\,s^{-1}}$
Wavevector	k	$\mathrm{m^{-1}}$
Attenuation vector	a	$\mathrm{m^{-1}}$
Dielectric constant	ε_r	dimensionless
Refractive index	n	dimensionless
Absorption index	κ	dimensionless
Absorption coefficient	α	$\mathrm{m^{-1}}$

References

[1] R. Regener and W. Sohler, "Loss in Low-Finesse Ti:$LiNbO_3$ Optical Waveguide Resonators", *Applied Physics B* **36**, 143–147 (1985).

[2] See, for instance, A. Yariv, *Quantum Electronics*, Chapter 11, John Wiley and Sons, Chichester, 1989.
[3] A.K. Ghatak and K. Thyagarajan, *Optical Electronics*, Cambridge University Press, Cambridge, 1989.
[4] R. Ulrich, "Theory of the Prism-Film Coupler by Plane-Wave Analysis", *Journal of the Optical Society of America*, **60**, 1337–1350 (1970).
[5] R.G. Heideman and P.V. Lambeck, "Remote Opto-Chemical Sensing with Extreme Sensitivity: Design, Fabrication and Performance of a Pigtailed Integrated Optical Phase-Modulated Mach-Zehnder Interferometer System", *Sensors and Actuators B* **61**, 100–127 (1999).

Further Reading

M. Born and E. Wolf, *Principles of Optics*, Pergamon Press, Oxford, 1980.
B.E.A. Saleh and M.C. Teich, *Fundamentals of Photonics*, John Wiley & Sons, Inc., Chichester, 1991.

3

THEORY OF INTEGRATED OPTIC WAVEGUIDES

Introduction

Integrated photonics devices are based on the processing of light confined in optical structures called *optical waveguides*. In this chapter we will describe the theory of optical waveguides using the electromagnetic theory of light previously discussed. We start by describing the basic geometries found in waveguide structures: planar waveguides, channel waveguides, optical fibres and photonic crystals. Then, we introduce the concept of optical mode, and discuss the types of modes that can be supported by a planar structure. Using the Maxwell's equations we will obtain the wave equation for planar waveguides, and then we will solve it for the simplest case of step-index planar waveguides, considering TE and TM polarised modes separately. In particular, we will see that for confined radiation the wave equation admits a discrete number of solutions, called *guided modes*. Based on the results obtained for guided modes in step-index waveguides, we describe the different approaches to solving graded-index planar waveguides, examining the advantages and disadvantages of each method. Finally, the guided modes in channel waveguides are studied, and the solutions provided by Marcatili's method and the effective index method are presented. Several examples of mode calculations are given for the different optical waveguide structures using the methods and algorithms described in the chapter.

3.1 Optical Waveguides: Basic Geometries

The basic element in integrated photonic technology is the optical waveguide. A waveguide can be defined as an optical structure that allows the confinement of light within its boundaries by total internal reflection. As we saw in the previous chapter, in order for total internal reflection to take place it is necessary to surround a high index medium, where most of the radiation energy is concentrated, by low refractive index media. A very simple example of light confinement happens in a (planar) film of glass situated in air. If the refractive index of the glass is n, the rays inside the film that propagate with an internal angle θ greater than the critical angle $\theta_c = \sin^{-1}(1/n)$ will suffer total reflection at the interfaces, and will remain trapped inside the film: in these circumstances we say that the film situated in air acts as an optical waveguide.

This is the basic mechanism that operates in *luminescent solar concentrators* (LSC) [1], that consist of a glass film in which some organic luminescent molecules (*dyes*) are embedded (Figure 3.1). The dye molecules absorb solar radiation, and then the radiation is re-emitted isotropically. An important fraction of the emitted radiation is trapped by total internal reflection at the upper and lower interfaces, and reaches the sides of the film, where a stack of solar cells are attached; in this way, a high geometrical collection factor is obtained.

Another type of optical waveguide, that can easily be visualised is a solid cylindrical glass tube (Figure 3.2): since the refractive index of glass is greater than the outer medium (air in this case), radiation travelling at angles greater than the critical angle will be confined in it by total internal reflection: in this case the confinement of the light extends in two dimensions.

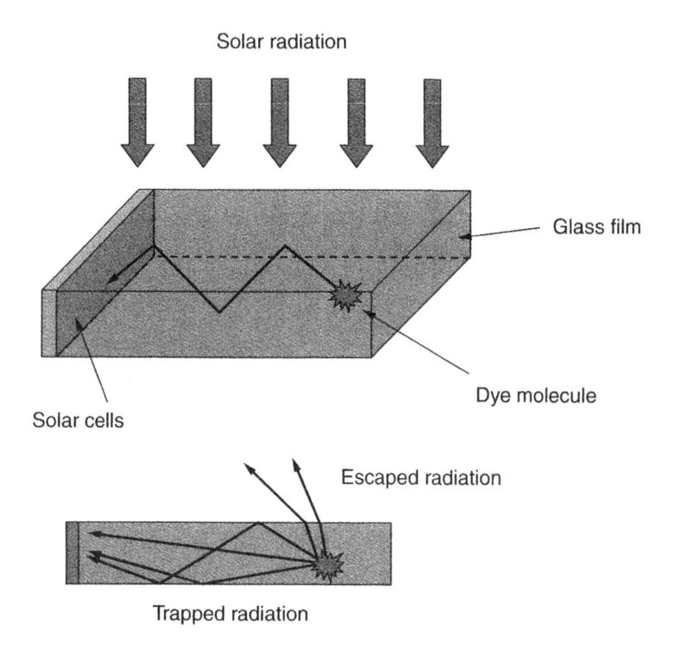

Figure 3.1 Luminescent solar concentrator acting as an optical waveguide: all the emitted radiation at angles greater than the critical angle are totally trapped in the dye-doped glass film, and reach the solar cell stack situated at the lateral face

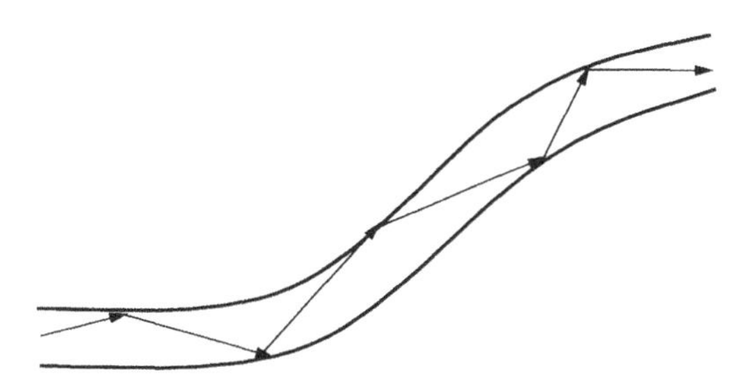

Figure 3.2 A solid glass tube can act as an optical pipeline by confining the light in two dimensions due to light undergoing total internal reflection at the glass–air interface

As the two former examples have shown, a first waveguide classification can be made by looking at the number of dimensions in which the light is confined (Table 3.1). Figure 3.3 shows the three basic types of waveguides depending on their number of dimensions for light confinement: while a planar waveguide (or 1D waveguide) confines the radiation in one dimension (Figure 3.3a), channel waveguides (or 2D waveguides) confine the light in two dimensions (Figure 3.3b).

There also exist structures that confine light in the three dimensions. These constitute a very special case of optical waveguides: since the radiation is confined in all directions, it cannot propagate. Therefore, these structures in fact form light traps, and are often called *photonic crystals* (Figure 3.3c). The light confinement in this case obviously cannot be based on total internal reflection; instead, photonic crystals are fabricated by means of tridimensional periodical structures, in which the light confinement is based on Bragg reflection. Photonic crystals have very interesting properties, and their use in several devices or applications has been proposed, such as miniaturised lasers with virtually no threshold power, waveguide bends with very small curvature radii and dimensions, or narrow-band filters [2].

Up to now, we have not imposed any restriction on the size of the guiding structures. In fact, the LSC and the light tubes can be called macroscopic waveguides. If we

Table 3.1 Classification of optical waveguides according to the number of dimensions of light confinement

Dimensions of light confinement	Classification of optical waveguides
1D	Planar waveguides
2D	Channel waveguides
	Optical fibres
3D	Photonics crystals

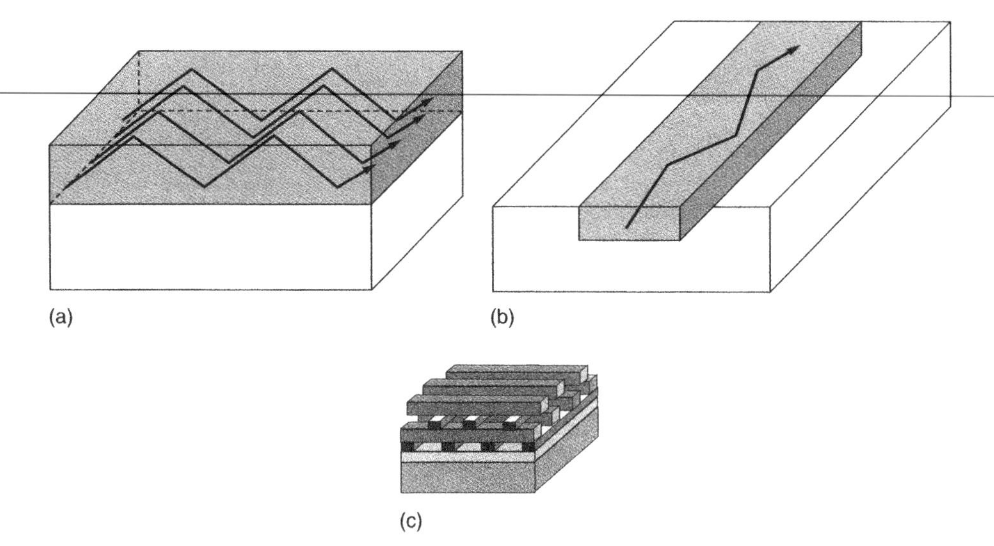

Figure 3.3 Basic types of waveguide geometries: (a) planar waveguide; (b) channel waveguide; (c) photonic crystal

start from a planar waveguide, and progressively reduce the thickness of the guiding film, when sizes of the order of the wavelength of the radiation are reached, a new phenomenon occurs: we found that due to the interference produced by the reflected wave coming from total internal reflection at the upper boundary and the reflected wave from the lower interface, now the light propagation is only allowed for a discrete set of angles. For each permitted angle of propagation, the transversal structure of the electromagnetic field associated with the radiation is maintained as the light beam propagates along the film; these characteristics form a *propagation mode*. Thus, a propagation mode is the result of combining total internal reflection and constructive interference.

The description that we have given to illustrate the concept of propagation modes was based on considering the light as plane waves, or in other words, that the direction of the light propagation within the waveguide can be described by using rays. In fact, although the ray optics treatment can give some interesting results, for a complete understanding and description of light propagation within guiding structures, it is necessary to consider light as electromagnetic waves, and use the formalism developed in the previous chapter.

Now we will describe the typical geometries found in optical waveguides, including planar waveguides, channel waveguides and optical fibres, and will discuss some of their basic characteristics.

Planar waveguides Planar waveguides are optical structures than confine optical radiation in a single dimension. Considering the refractive index distribution in the planar structure, planar waveguides can be classified as *step-index waveguides* or *graded index waveguides*.

The step-index planar waveguide is the simplest structure for light confinement, and is formed by a uniform planar film with a constant refractive index (homogeneous film, n_f = constant), surrounded by two dielectric media of lower refractive indices [3]. The homogeneous upper medium, or *cover*, has a refractive index of n_c, and the lower medium, with refractive index n_s, is often called *substrate*. Usually, it is assumed that the refractive index of the cover is less than or equal to the refractive index of the substrate, $n_c \leq n_s$, and in this way we have $n_f < n_s \leq n_c$. In fact, in many cases the cover medium is air, and therefore $n_c = 1$, which fulfils the assumption previously mentioned.

If the upper and the lower media are the same (equal optical constants), the structure forms a *symmetric planar waveguide*. On the other hand, in integrated photonics the upper and lower media are different, and in this case we are dealing with an *asymmetric planar waveguide* (Figure 3.4). Asymmetric step-index planar waveguides are fabricated by depositing a high-index film on top of a lower index substrate, by means of physical methods (thermal evaporation, molecular beam epitaxy, sputtering, etc.) or chemical methods (chemical vapour deposition, metal-organic chemical vapour deposition, etc.).

If the high index film is not homogeneous, but its refractive index is depth dependent (along the x-axis in Figure 3.5), the structure is called a *graded index planar waveguide* [4]. Usually the refractive index is maximum at the top surface, and its value decreases with depth until it reaches the value corresponding to the refractive index of the substrate (Figure 3.5). This kind of structure is present in waveguide fabrication methods based on the surface modification of a substrate, whether by physical

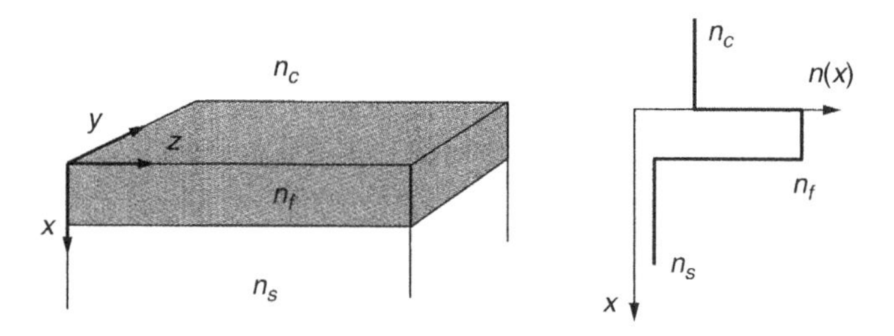

Figure 3.4 Asymmetric step index planar waveguide. Right: refractive index profile, where $n_f < n_s \leq n_c$

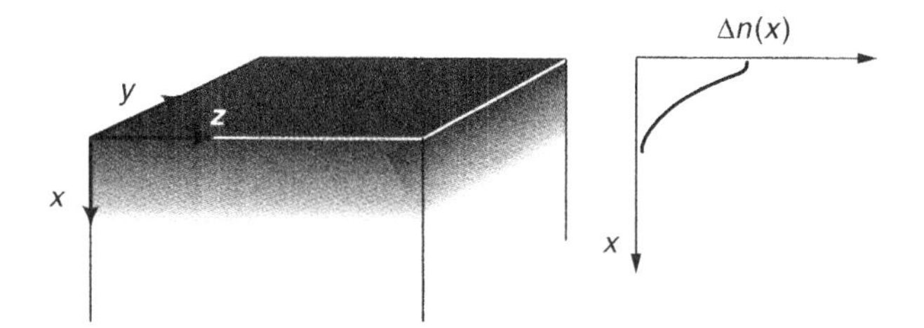

Figure 3.5 Graded index planar waveguide

processes (ion implantation, metal diffusion, etc.), or by chemical modification of the substrate (ionic exchange methods).

Channel waveguides In planar waveguides the light confinement is restricted to a single dimension (the x-axis in Figures 3.4 and 3.5), and if the light propagates along a given direction (z-axis), the light can spread out in a perpendicular direction (y-axis) due to diffraction. When we want to avoid this effect and keep the light beam well confined, it is necessary for total internal reflection to take place not only at the upper and lower interfaces, but also at the lateral boundaries. This confinement is attained in channel waveguides, or 2D waveguides, in which the core region (where the radiation is concentrated) has a refractive index greater than any of the surrounding media (Figure 3.3b). The classification made for planar waveguides, in terms of symmetric/asymmetric or step index/graded index, is also valid for channel waveguides, but with the difference that we are now dealing with the extra dimension which characterises the waveguide structure.

Although many types of channel waveguides have been proposed, three are the most common basic structures used. The easiest way to build a channel waveguide is to deposit a stripe made of a high refractive index material on top of a lower refractive index substrate. This kind of channel waveguide is called *stripe waveguide*, and can be made by either depositing the stripe directly onto the substrate, or simply by conveniently etching a previously deposited film (Figure 3.6a) [5]. If the etch process is not complete and does not reach the substrate, a channel waveguide is also produced, providing that the thickness and height of the structure are conveniently tailored; this waveguide geometry is called *rib waveguide* (Figure 3.6b). Another type of channel waveguide common in integrated photonic is the *buried channel waveguide*

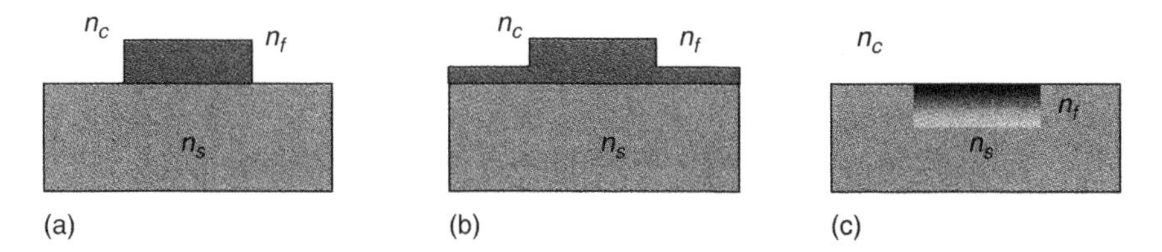

Figure 3.6 Three main types of channel waveguides (2D waveguides): (a) stripe waveguide; (b) rib waveguide; (c) buried waveguide

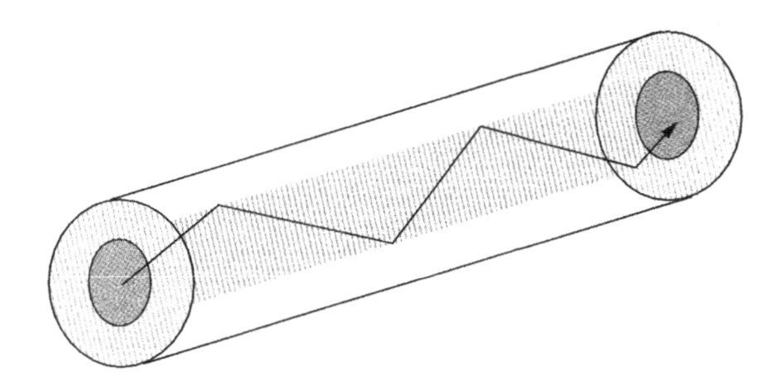

Figure 3.7 Optical path followed by a ray of light inside an optical fibre. The light is confined in two dimensions due to the total internal reflection occurring at the core–cover interface

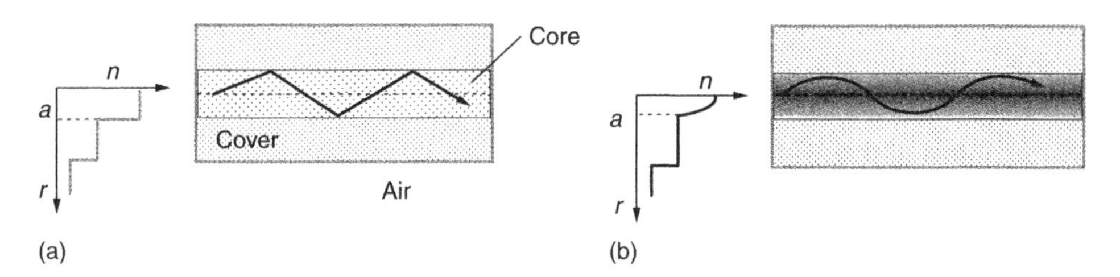

Figure 3.8 Radial dependence on the refractive index in an optical fibre with core radius a: (a) step-index optical fibre; (b) graded index optical fibre

(Figure 3.6c). In this case the waveguide is fabricated by inducing a local increase of the substrate's refractive index, which is usually performed by diffusion methods [6]. Of course, the channel waveguides, either stripe or buried, that are the basic building blocks of integrated photonic devices, require lithographic techniques, as was discussed in Chapter 1.

Optical fibres A special type of channel waveguides, from the point of view of their geometry and manufacturing methods as well as their applications, is called optical fibres (Figure 3.7) [7]. Optical fibres have cylindrical geometry, and are constituted by a cylindrical *core* of radius a and refractive index n_1, surrounded by a cladding of slightly lower refractive index n_2. As we did with planar waveguides, we can classify the optical fibres as step-index fibres, with homogeneous core (n_1 = constant, Figure 3.8a), or graded index fibres, in which the refractive index of the core varies as a function of the radial distance ($n_1 = n_1(r)$, Figure 3.8b). This last type of optical fibres is the best choice when high transmission bandwidth is required in long-distance optical communications [7].

3.2 Types of Modes in Planar Optical Waveguides

Light behaviour in an optical waveguide can initially be analysed by examining the case of an asymmetric planar waveguide from the point of view of geometric optics (ray optics).

Let us consider the planar waveguide depicted in Figure 3.9, where we have assumed that the refractive index of the film n_f is higher than the refractive index corresponding to the substrate n_s and the upper cover n_c. In addition, we assume the usual situation in which the relation $n_s > n_c$ is fulfilled. In this way, the critical angles that define total internal reflection for the cover–film interface (θ_{1c}) and the film–substrate boundary (θ_{2c}) are determined by:

$$\theta_{1c} = \sin^{-1}(n_c/n_f) \tag{3.1}$$

$$\theta_{2c} = \sin^{-1}(n_s/n_f) \tag{3.2}$$

In addition, as we have $n_f > n_s > n_c$, it follows that the critical angles fulfil the relation $\theta_{2c} > \theta_{1c}$. If now we fix our attention to the propagating angle θ of the light inside the film (Figure 3.9), three situations can be distinguished:

(i) $\theta < \theta_{1c}$. In this case, if the ray propagates with internal angles θ lower than the critical angle corresponding to the film–cover interface θ_{1c}, the light penetrates the cover, as well as the substrate, because $\theta_{2c} > \theta_{1c}$. Thus, the radiation is not confined to the film, but travels in the three regions. This situation corresponds to *radiation modes*, because the light *radiates* to the cover and the substrate (Figure 3.10).

(ii) $\theta_{1c} < \theta < \theta_{2c}$. Light travelling in these circumstances is totally reflected at the film–cover interface, thus it cannot penetrate the cover region. Nevertheless, the radiation can still penetrate the substrate, and therefore it corresponds to *substrate radiation modes*, or in short, *substrate modes* (Figure 3.11).

(iii) $\theta_{2c} < \theta < \pi/2$. In this situation, the ray will suffer total internal reflection at the upper and lower interfaces, and thus the radiation is totally confined and cannot escape the film. This situation corresponds to a guided mode (Figure 3.12), and is the most relevant case in integrated optics.

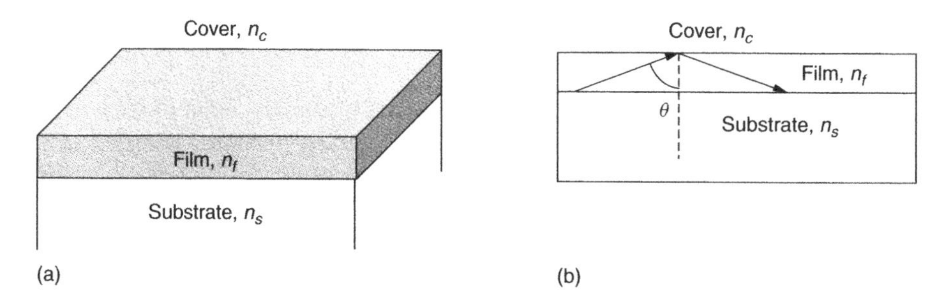

Figure 3.9 (a) Asymmetric planar waveguide. (b) Zig-zag trajectory of a ray inside the film

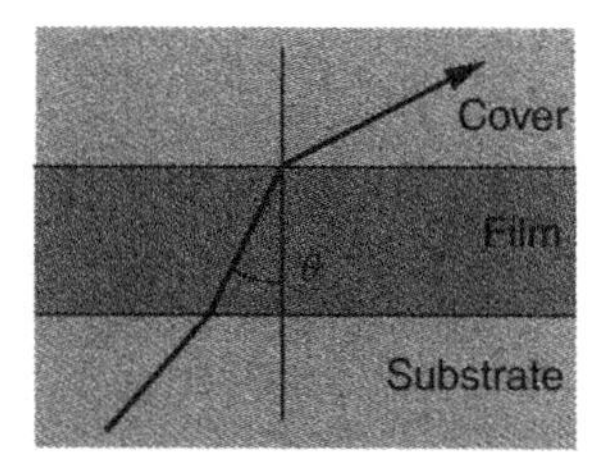

Figure 3.10 Radiation mode in an asymmetric step-index planar waveguide

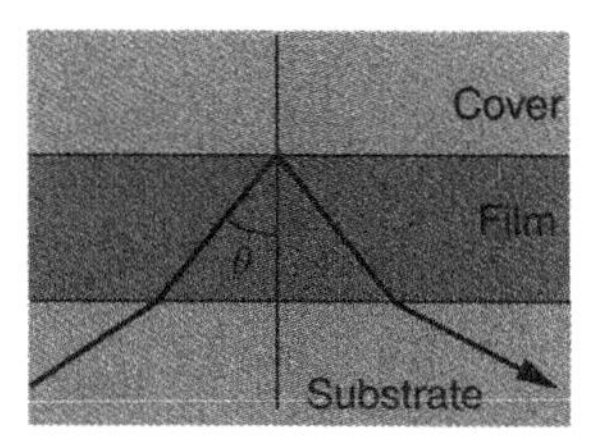

Figure 3.11 Ray path followed by a substrate radiation mode

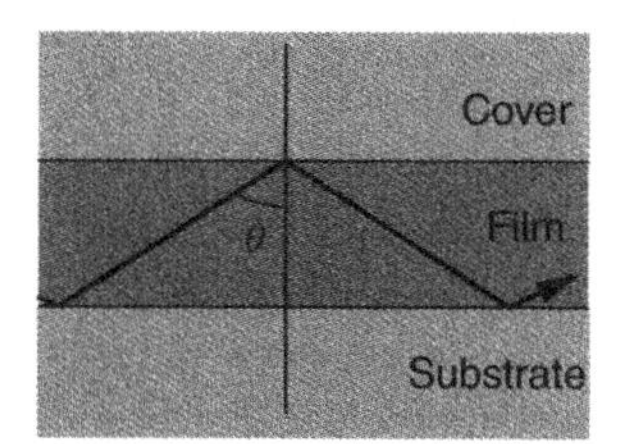

Figure 3.12 Guided mode in an asymmetric planar waveguide, showing the zig-zag path traced by the ray

Ray optics analysis of guided modes Although the light propagation in waveguide structures should be analysed by a rigorous electromagnetic wave treatment, an analysis based on optical rays is not only more intuitive, but in addition its solution to the problem coincides with that supplied by the more rigorous wave treatment. The optical ray approach of the guided modes in planar optical waveguides consists of studying a ray inside the film moving on a zig-zag path. The first condition which a ray of light must fulfil in order to be confined in the film region is that the angle of incidence at the upper and lower interfaces must be higher that the critical angles defined by the cover–film and film–substrate boundaries (Figure 3.13), that is, $\theta < \theta_{1c}, \theta_{2c}$.

In a round trip inside the film, the ray suffers a transversal phase shift that depends on the film thickness, and also additional phase shifts due to total internal reflection at the two boundaries. The condition for a guided mode is established on the basis of constructive interference, which implies that the total transversal phase shift in a complete round trip should be an integral number of 2π. Only a discrete number of angles fulfils that condition, and these will correspond to the propagation angles of guided modes.

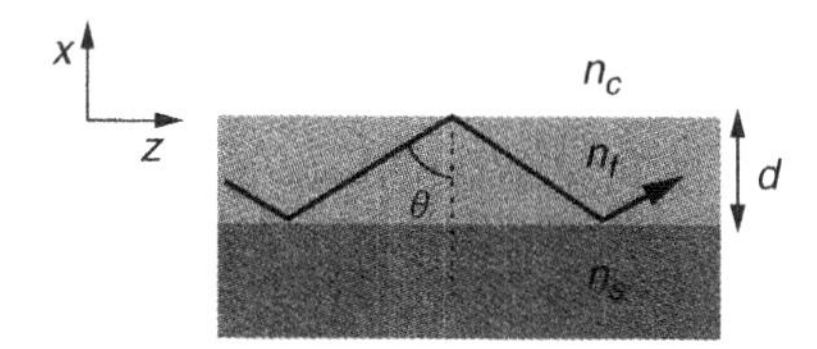

Figure 3.13 Ray tracing a zig-zag path in an asymmetric step-index planar waveguide

The transversal phase shift experimented by the ray due to the zig-zag path in a round trip is given by:

$$\phi = 2k_x d = 2k_0 n_f d \cos\theta \tag{3.3}$$

where k_x is the x-component of the wavevector. On the other hand, the phase shifts suffered by the ray due to total internal reflection at the upper (ϕ_c) and lower boundaries (ϕ_s) are given by equations (2.148) and (2.149), for TM and TE polarisations respectively. The self-consistency condition for constructive interference implies that the total phase shift must be zero or an integer multiple of 2π:

$$2k_0 n_f d \cos\theta - \phi_c - \phi_s = 2\pi m \tag{3.4}$$

where the integer m is the *mode order*. This relation is known as the *transversal resonance condition* for guided modes in asymmetric step-index planar waveguides, and it is a transcendental equation involving waveguide parameters (n_c, n_f, n_s and d), the working wavelength λ and the propagation angle θ. For a particular mode order m, the equation (3.4) can be solved numerically, and the modal angle is obtained.

The *propagation constant* of the mode β_m is the component of the wavevector along the propagation direction, thus the relation with the propagation angle is:

$$\beta_m = k_0 n_f \sin\theta_m \tag{3.5}$$

This constitutes a link between the ray picture of the guided modes, characterised by its propagation angle θ_m, and the electromagnetic wave treatment that considers the mode characterised by its propagation constant β_m.

The ray optic approach that we have carried out can be used for the qualitative description of light behaviour in an optical waveguide, to establish the types of mode that can be found in such structures, to calculate the number of guided modes that support a waveguide, and to determine its propagation constants. Nevertheless, for many applications it is essential to know the electric field distribution of the radiation within the waveguide structure, and this method does not provide such information. If one wants to determine the optical fields or the intensity distribution associated with the light propagation in waveguide structures, it becomes necessary to invoke a more rigorous formalism, based on the electromagnetic theory of the light, as explained in Chapter 2. Therefore, the problem should start from Maxwell's equations applied to the electromagnetic fields in a given structure, which defines the waveguide; the solutions for the fields will correspond to the propagation modes.

3.3 Wave Equation in Planar Waveguides

We will now discuss the electromagnetic theory of light applied to a planar waveguide, because it is the simplest structure to be analysed from the point of view of its mathematical description, and from it the general features related to more complex waveguide geometry can be understood. Starting from Maxwell's equations and from the constitutive relations, we will obtain the wave equations for TE and TM propagation that govern light behaviour in planar waveguides. These wave equations will be solved for the general case of asymmetric step-index planar waveguides, and later on we will discuss some methods for solving the wave equations in more complex planar waveguides, as is the case of graded-index planar waveguides. Finally, we will discuss the problem in the modelling of channel waveguides, and examine some approximate methods that can be applied to calculate the propagation modes in 2D structures, such as the *effective index method* and *Marcatili's method.*

Assuming that the light is propagating through a dielectric (conductivity $\sigma = 0$), non-magnetic (magnetic permeability $\mu = \mu_0$), isotropic and linear medium ($\mathcal{D} = \varepsilon\mathcal{E}$), Maxwell's equations are reduced to:

$$\nabla x\mathcal{E} = -\mu_0 \frac{\partial \mathcal{H}}{\partial t} \tag{3.6}$$

$$\nabla x\mathcal{H} = \varepsilon_0 n^2 \frac{\partial \mathcal{E}}{\partial t} \tag{3.7}$$

where $\mathcal{E}$ and $\mathcal{H}$ are the electric and magnetic fields respectively, μ_0 is the free space permeability, ε_0 is the permittivity of the free space and n is the refractive index of the medium where the light propagates.

If the medium is optically inhomogeneous its properties are position-dependent, in particular the refractive index, $n = n(\boldsymbol{r})$. From Maxwell's equations (3.6) and (3.7) the following wave equations for $\mathcal{E}$ and $\mathcal{H}$ are derived:

$$\nabla^2\mathcal{E} + \nabla\left(\frac{1}{n^2}\nabla n^2\mathcal{E}\right) - \varepsilon_0\mu_0 n^2 \frac{\partial^2\mathcal{E}}{\partial t^2} = 0 \tag{3.8}$$

$$\nabla^2\mathcal{H} + \frac{1}{n^2}\nabla n^2 x(\nabla x\mathcal{H}) - \varepsilon_0\mu_0 n^2 \frac{\partial^2\mathcal{H}}{\partial t^2} = 0 \tag{3.9}$$

These two vectorial equations indicate that for an inhomogeneous medium the Cartesian components of the electric field vector $\mathcal{E}_x$, $\mathcal{E}_y$ and $\mathcal{E}_z$ (and the components of the magnetic vector) are coupled, and therefore we cannot establish a scalar equation for each component as we did in the case of a homogeneous medium. Only in light propagation in a homogeneous medium, in which the refractive index is constant ($\nabla n^2 = 0$), the second terms in equations (3.8) and (3.9) vanish, and each of the Cartesian components for the fields $\mathcal{E}$ and $\mathcal{H}$ satisfy the scalar wave equation (2.24).

If the refractive index of the inhomogeneous medium depends only on two Cartesian coordinates, for instance x and y, so that $n = n(x, y)$, and we choose the third coordinate (z) as the propagation direction of the radiation, the solutions for the inhomogeneous wave equations (3.8) and (3.9) for monochromatic waves can be written as:

$$\mathcal{E}(\mathbf{r}, t) = \mathbf{E}(x, y)e^{i(\omega t - \beta z)} \tag{3.10}$$

$$\mathcal{H}(\mathbf{r}, t) = \mathbf{H}(x, y)e^{i(\omega t - \beta z)} \tag{3.11}$$

ω being the angular frequency and β the propagation constant of the wave. These two expressions determine the electromagnetic field for a propagating mode, which is characterised by its propagation constant β. This solution is found in light propagation in straight channel waveguides or in optical fibres, because in both cases the structure, defined by the spatial dependence of the refractive index, is invariant with the z-coordinate.

Assuming now that the refractive index depends only on a single Cartesian coordinate, for instance $n = n(x)$, which is the case of planar optical waveguides, the spatial part of the complex exponential function in the expressions (3.10) and (3.11) takes the form $-i(\gamma y + \beta z)$. If we further assume propagation along the z-axis, the wave has no dependence on the y-axis, thus $\gamma = 0$, and the electric and magnetic fields take the form:

$$\mathcal{E}(\mathbf{r}, t) = \mathbf{E}(x)e^{i(\omega t - \beta z)} \tag{3.12}$$

$$\mathcal{H}(\mathbf{r}, t) = \mathbf{H}(x)e^{i(\omega t - \beta z)} \tag{3.13}$$

Therefore, given a refractive index distribution $n(x)$ that defines the planar waveguide, the solutions for the electromagnetic fields that support that waveguide are reduced to find out the solutions for the complex field amplitudes $\mathbf{E}(x)$ and $\mathbf{H}(x)$ as well as for the propagation constants β. We will show that for a particular propagation constant β, whether corresponding to a confined mode or a radiation mode, the field distributions are completely determined. Thus, providing that the polarisation character of the light has been initially established, a mode is one-to-one defined by its propagation constant.

In order to find the propagation modes in a planar waveguide we will study two independent situations: in the first case, the electric field associated with the mode has only a transversal component, and so its solutions are the *TE modes*; the second case involves the situation in which the electric field has only a parallel component, and the solutions are called *TM modes*.

TE modes In this case we must find the general solution for the complex amplitudes $\mathbf{E}(x)$ and $\mathbf{H}(x)$ when the electric field vector has only perpendicular component (referred to the incident plane, as discussed in Section 2.2). Following the geometry on Figure 3.14, the perpendicular component of the electric field corresponds to E_x, and thus $E_y = E_z = 0$. On the other hand, the magnetic field satisfies $H_y = 0$.

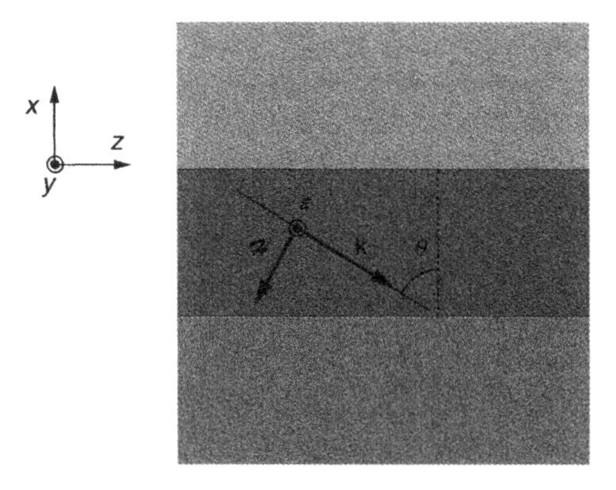

Figure 3.14 TE propagation in an asymmetric planar waveguide

Therefore, the only components of the electric and magnetic fields are E_y, H_x and H_z. By substituting the solution for planar structures given by equations (3.12) and (3.13) in Maxwell's equations (3.6) and (3.7), we obtain the following coupled equations:

$$H_x = -(\beta/\omega\mu_0)E_y \tag{3.14}$$

$$H_z = (i/\omega\mu_0)\partial E_y/\partial x \tag{3.15}$$

$$i\beta H_x + \partial H_z/\partial x = -i\omega\varepsilon_0 n^2(x)E_y \tag{3.16}$$

that relate the field amplitude components E_y, H_x and H_z.

If now we substitute the H_x and H_z components from equations (3.14) and (3.15) respectively into equation (3.16), and taking into account that the partial derivatives are in this case total derivatives because the electric field component depends only on the x coordinate, we obtain an equation involving only the E_y component of the electric field:

$$\frac{d^2E_y(x)}{dx^2} + \left[k_0^2n^2(x) - \beta^2\right]E_y(x) = 0 \tag{3.17}$$

This differential equation is the wave equation that must satisfy the electric field amplitude $E_y(x)$ for TE propagation in planar structures, where $k_0 = 2\pi/\lambda_0$, and λ_0 is the wavelength of the light in free space, related to the angular frequency by $\omega = 2\pi c/\lambda_0$. Since the wave equation (3.17) is a second-order differential equation, in order to resolve it for a given planar structure, it will be necessary to impose additional conditions. Indeed, we must impose the appropriate boundary conditions at the interfaces, as we showed in the previous chapter. In particular, we should invoke the continuity of the tangential components of the electric field $\boldsymbol{\mathcal{E}}_t$ and magnetic field $\boldsymbol{\mathcal{H}}_t$ at the interfaces. In the case of TE propagation through an asymmetric planar waveguide, these boundary conditions imply the continuity of the E_y and H_z at the cover–film interface and at the substrate–film boundary.

From equation (3.15) we have that the z component of the magnetic field H_z is proportional to the first derivative of the electric field component E_y, and so the boundary conditions lead to the continuity of the electric field component E_y besides the continuity of its first derivative dE_y/dx.

TM modes Let us now consider the case of light propagation in which its associated electric field vector has only a parallel component to the incident plane, so the

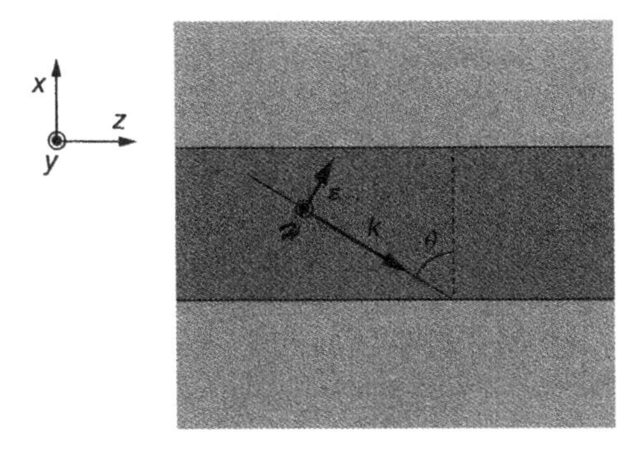

Figure 3.15 TM propagation in an asymmetric planar waveguide

only component of the magnetic field is perpendicular to that plane, as indicated in Figure 3.15.

For TM light propagation, and following Figure 3.15, the non-vanishing components of the electric and magnetic field vectors are now E_x, E_z and H_y. In a similar way to that undertaken for TE propagation, from Maxwell's equations (3.6) and (3.7), and using a solution such as that postulated by (3.12) and (3.13), we obtain the following relation between the field components for TM propagation:

$$E_x = (\beta/\omega\varepsilon_0 n^2)H_y \tag{3.18}$$

$$E_z = (1/i\omega\varepsilon_0 n^2)\partial H_y/\partial x \tag{3.19}$$

$$i\beta E_x + \partial E_z/\partial x = i\omega\mu_0 H_y \tag{3.20}$$

Since the magnetic field vector has now a single non-vanishing component (H_y), we will use it to establish the wave equation for TM modes. By substituting E_x and E_z from equations (3.18) and (3.19) respectively into equation (3.20), it yields:

$$\frac{d^2H_y(x)}{dx^2} - \frac{1}{n^2}\frac{dn^2}{dx}\frac{dH_y(x)}{dx} + \left[k_0^2n^2(x) - \beta^2\right]H_y(x) = 0 \tag{3.21}$$

This is the wave equation for TM modes in planar waveguides, which is slightly more complex than the one found for TE modes because it contains an additional term that involves the product of the first derivatives of the magnetic field component dH_y/dx and the first derivative of the dielectric constant dn^2/dx. Nevertheless, in a constant refractive index region ($n = constant$), this additional term vanishes, and equation (3.21) simplifies to:

$$\frac{d^2H_y(x)}{dx^2} + \left[k_0^2n^2(x) - \beta^2\right]H_y(x) = 0 \tag{3.22}$$

This wave equation, valid for TM propagation modes in planar waveguides, is identical to that obtained for TE modes, with the exception that now the differential equation has been established as a function of the magnetic field instead of the electric field.

Once again, the complete solution of equation (3.22) requires us to impose the adequate boundary conditions at the interfaces. In this case, the continuity of the transversal field components of the electric and magnetic fields leads to the continuity of E_z and H_y field components at the interfaces. As E_z is related to H_y through equation (3.19), the continuity across the boundary must be fulfilled for H_y and $(1/n^2)dH_y/dx$.

At this point it is necessary to recall that Figures 3.14 and 3.15, used for defining TE and TM propagation modes, include not only the electric and magnetic field vectors, but we also have drawn a ray, with a very precise propagation angle and even a wavevector **k**. Although, strictly speaking, this picture is not physically correct because a wavevector can be defined only for monochromatic plane waves, the fact of introducing a ray can serve as a link between the electromagnetic treatment and the geometrical optic treatment. Indeed, the propagation angle θ can be obtained by considering the modes as plane waves, and the result agrees with the rigorous EM analysis using the relation $\beta = k_0 n_f \sin\theta$.

General solution of the wave equation for planar waveguides Once the wave equations for TE and TM modes have been established for planar waveguides, we will examine in particular what kind of solution admits the wave equation for TE polarisation, see equation (3.17). The solution in the case of light propagation with TM polarisation is basically the same, with the exception that the boundary conditions are slightly different, because of the factor $(1/n^2)$ in the continuity of the magnetic field component derivative.

Starting from the wave equation that must satisfy a TE mode that propagates in a planar waveguide, see equation (3.17), characterised by its propagation constant β, we postulate a solution for the $E_y(x)$ component in the form:

$$E_j(x) = A_j e^{i\gamma_j x} + B_j e^{-i\gamma_j x} \tag{3.23}$$

where $E_j(x)$ represents the y-component of the electric field amplitude for the wave propagating in the jth region, being the coefficients A_j and B_j two complex constants, which will be calculated after imposing the appropriate boundary conditions. It can be proved that expression (3.23) satisfies the wave equation (3.17) providing that the γ_j parameter is given by:

$$\gamma_j = \sqrt{k_0^2 n_j^2 - \beta^2} \tag{3.24}$$

where γ_j is different for each region, characterised by its refractive index n_j.

In order to discuss the different behaviours of the solution given by (3.23) in a particular region, it is convenient to introduce a new adimensional parameter, called *effective refractive index* N, directly related to the propagation constant β of the mode through the formula:

$$\beta \equiv k_0 N \tag{3.25}$$

Following this definition, the effective refractive index N represents the refractive index experienced by the mode propagating along the z-axis.

We shall now discuss the general solutions given by equation (3.23) depending on the effective refractive index N of the mode and the refractive index n_j of the region considered (Figure 3.16).

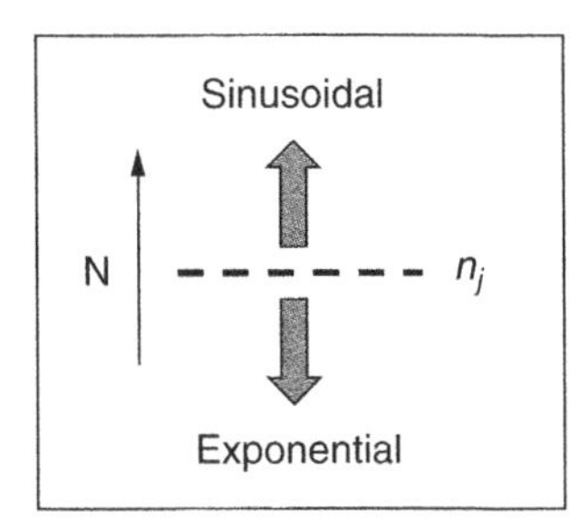

Figure 3.16 Behaviour of the solution for the electric field component in a given region depending on the effective refractive index of the propagation mode

(i) If the propagation constant β is lower than $k_0 n_j$ (or $N < n_j$) then the parameter γ_j is a real number, following the definition given in (3.24), and the general solution postulated by (3.23) will correspond to a sinusoidal function.
(ii) By contrast, if the propagation constant satisfies that $\beta > k_0 n_j$ (or $N > n_j$), the parameter γ_j is a pure imaginary number, and therefore the solution given by (3.23) should be described by exponential functions.

3.4 Guided Modes in Step-index Planar Waveguides

The general solution discussed in the previous paragraph can easily be applied to the case of guided modes supported by asymmetric step-index planar waveguides, considering the geometry given in Figure 3.17. The three media have refractive indices n_c (cover), n_f (film) and n_s (substrate), and are separated by planar boundaries perpendicular to the x-axis, the light propagation being along the z-axis. We further assume that $n_f < n_s < n_c$, and that the plane $x = 0$ corresponds to the cover–film boundary. Therefore, if the film thickness is d, the film–substrate interface is located at the plane $x = -d$.

Guided TE-modes Although step-index planar waveguides are structures inherently inhomogeneous, within each of the three region the refractive index is constant. Thus, considering each region separately, the wave equation for TE modes is expressed as:

$$\frac{d^2 E_y}{dx^2} + \left[k_0^2 n^2 - \beta^2\right] E_y = 0 \tag{3.26}$$

If we are interested in the search of guided modes, as we have seen before, the propagation constant β associated with a particular mode must fulfil the condition:

$$k_0 n_s < \beta < k_0 n_f \tag{3.27}$$

or in terms of the refractive indices, the effective refractive index N of the guided mode must be between the refractive index on the film n_f and the refractive of the substrate n_s (see Figure 3.18):

$$n_s < N < n_f \tag{3.28}$$

Bearing in mind this result, the wave equation (3.17) in each homogeneous region can be written as:

$$d^2 E_y/dx^2 - \gamma_c^2 E_y = 0 \quad x \geq 0 \quad \text{(Cover)} \tag{3.29}$$

$$d^2 E_y/dx^2 + \kappa_f^2 E_y = 0 \quad 0 > x > -d \quad \text{(Film)} \tag{3.30}$$

$$d^2 E_y/dx^2 - \gamma_s^2 E_y = 0 \quad x \leq -d \quad \text{(Substrate)} \tag{3.31}$$

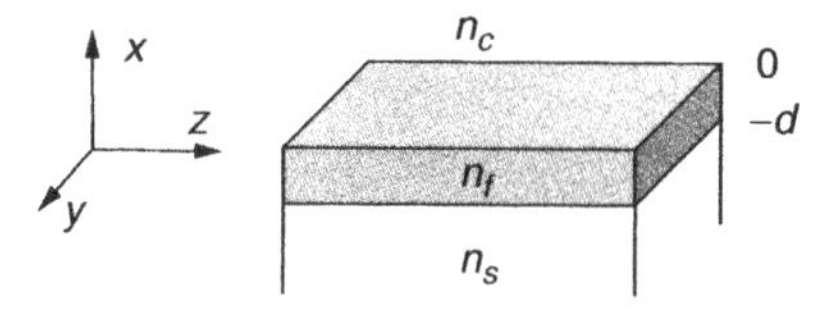

Figure 3.17 Geometry used for the analysis of propagating modes in an asymmetric step-index planar waveguide

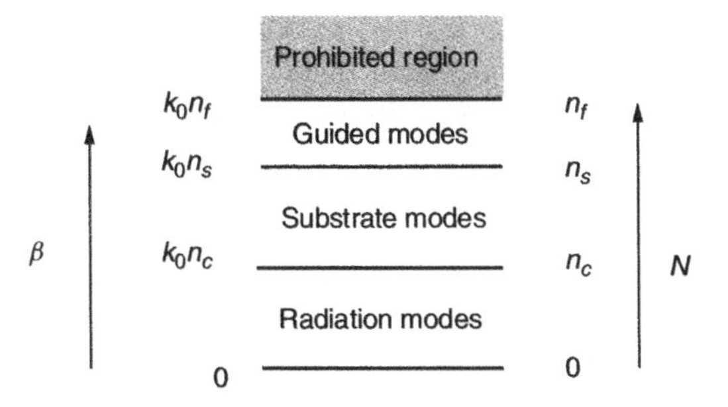

Figure 3.18 Range of values for the propagation constant β and the effective refractive index N for guided modes, substrate modes and radiation modes

where the three parameters γ_c, κ_f and γ_s are given by:

$$\gamma_c^2 = \beta^2 - k_0^2 n_c^2 \tag{3.32}$$

$$\kappa_f^2 = k_0^2 n_f^2 - \beta^2 \tag{3.33}$$

$$\gamma_s^2 = \beta^2 - k_0^2 n_s^2 \tag{3.34}$$

Considering these definitions and the range of β for confined modes, it is clear that γ_c, κ_f and γ_s are real numbers, and we take them as positive values.

By solving the differential equations (3.29)–(3.31), the electric field in the cover, film and substrate regions can be expressed as:

$$E_y = \begin{cases} Ae^{-\gamma_c x} & x \geq 0 \\ Be^{i\kappa_f x} + Ce^{-i\kappa_f x} & -d < x < 0 \\ De^{\gamma_s x} & x \leq -d \end{cases} \tag{3.35}$$

The electric field in the cover also admits an additional solution of the form of $A' exp(\gamma_c x)$, but as an increasing exponential function for $x > 0$ has not physical meaning for a confined mode, we have $A' = 0$. A similar reasoning has been used to eliminate the term $D' exp(-\gamma_s x)$ corresponding to the substrate region.

The boundary conditions require that E_y and dE_y/dx must be continuous at the cover–film interface ($x = 0$) and at the film–substrate frontier ($x = -d$), giving place to four equations that relate the constant parameters A, B, C and D and the propagation constant β. Therefore, we have five unknown quantities to be determined from only a set of four equations. Indeed, one of the constant parameters cannot be determined and should remain free (for instance, the parameter A), and it will be determined once the energy carried by the propagating mode is settled. By solving this set of equations, and after cumbersome calculation, the following equation is obtained:

$$\tan \kappa_f d = \frac{\dfrac{\gamma_c}{\kappa_f} + \dfrac{\gamma_s}{\kappa_f}}{1 - \left(\dfrac{\gamma_c}{\kappa_f}\right)\left(\dfrac{\gamma_s}{\kappa_f}\right)} \tag{3.36}$$

This relation can be considered as the *dispersion relation* for the asymmetric step-index planar waveguide, and is a transcendental equation involving the parameters that define the waveguide structure (n_c, n_f, n_s and d), the working wavelength λ and the propagation constant β of the guided mode, and from which one can calculate numerically the propagation constant β. In addition, as the tangent function fulfil:

$$\tan(\kappa_f d) = \tan(\kappa_f d + m\pi) \quad m = 0, 1, 2, \ldots \tag{3.37}$$

in general there exist several solutions for the propagation constant β depending on the integer number m. This integer number m is called the *mode order*, and the associated propagation constant is referred as β_m.

It is convenient to define a set of parameters, called *normalised parameters*, in such a way that the transcendental equation (3.37) can be universalised for any asymmetric step-index waveguide. These parameters are defined as:

$$b = (N^2 - n_s^2)/(n_f^2 - n_s^2) \quad \text{Normalised mode index} \tag{3.38}$$

$$V = k_0 d (n_f^2 - n_s^2)^{1/2} \quad \text{Normalised film thickness} \tag{3.39}$$

$$a = (n_s^2 - n_c^2)/(n_f^2 - n_s^2) \quad \text{Asymmetry measure} \tag{3.40}$$

As the effective refractive index corresponding to a confined mode is in the range $n_s < N < n_f$, the normalised mode index b is bounded to $0 < b < 1$. On the other hand, the normalised film thickness V is directly connected to the relative thickness of the waveguide core (film) with respect to the working wavelength, that is, $V \propto d/\lambda$, as deduced from relation (3.39). Finally, the asymmetry measure a is zero in the case of symmetric waveguides, and increases as the refractive index difference between the cover and substrate increases.

The transcendental equation (3.36), written as function of the propagation constant β, is rewritten in terms of the normalised parameters as:

$$\tan\left[V\sqrt{1-b}\right] = \frac{\sqrt{\dfrac{b}{1-b}} + \sqrt{\dfrac{b+a}{1-b}}}{1 - \dfrac{\sqrt{b(b+a)}}{(1-b)}} \tag{3.41}$$

In general, equations (3.36) or (3.41) admit a finite number of solutions for a finite number of the integer m, and thus the waveguide will support a finite number of guided modes. In this case, we refer to it as a *multi-mode waveguide*. In the particular case in which the dispersion equation only admits a solution for $m = 0$, the waveguide is called a *monomode waveguide*. Moreover, it is possible that a given structure has no solution for the transcendental equation (3.36), and in this case (for a particular working wavelength) the waveguide cannot support any guided mode.

Figure 3.19 shows the numerical solution of the dispersion equation (3.41) for a symmetric ($a = 0$) and asymmetric ($a = 50$) waveguide as a function of their normalised parameters b and V, where we have included the solution for the mode orders $m = 0$, $m = 1$ and $m = 2$. For example, a symmetric waveguide characterised by $V = 4$ will support two TE modes ($m = 0$, 1); in contrast, for the same normalised film thickness ($V = 4$) an asymmetric waveguide with $a = 50$ will only admit a confined TE mode ($m = 0$).

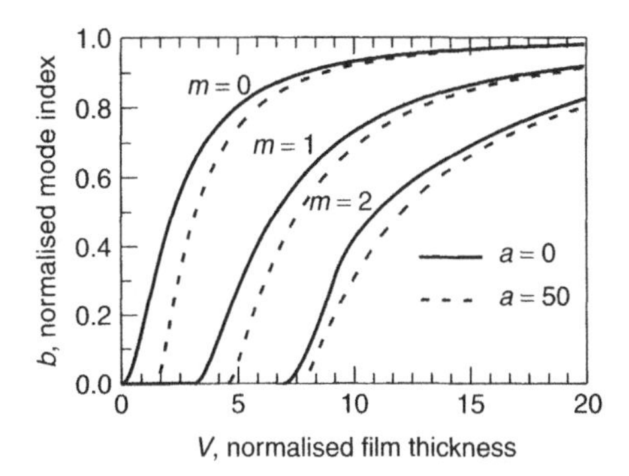

Figure 3.19 Dispersion relations for a symmetric step-index planar waveguide and for an asymmetric waveguide ($a = 50$), as a function of their normalised parameters b and V

Once the propagation constant β (or b) of a mode has been calculated, the coefficients γ_c, κ_f and γ_s are defined straightforwardly, and thus the electric field in the three regions can now be completely determined:

$$E_y(x) = \begin{cases} Ae^{-\gamma_c x} & x \geq 0 \\ A\left(\cos\kappa_f x - \dfrac{\gamma_c}{\kappa_f}\sin\kappa_f x\right) & 0 < x < -d \\ A\left(\cos\kappa_f d + \dfrac{\gamma_c}{\kappa_f}\sin\kappa_f d\right)e^{\gamma_s(x+d)} & x \leq -d \end{cases} \quad (3.42)$$

According to this expression, the electric field decreases exponentially in the cover and in the substrate while its dependence is sinusoidal in the film, as was expected for the behaviour of a confined mode. Figure 3.20 shows the electric field profiles for the four confined modes $m = 0$, 1, 2, and 3, supported by a planar waveguide formed by a 3 μm thick film of refractive index 1.50, surrounded by air and by a substrate of refractive index 1.43. These modes have been calculated for a wavelength of $\lambda = 0.633$ μm. As can be observed from Figure 3.20, the electric field as well as its derivative are continuous at both interfaces. The solution for E_y, nevertheless, is completely determined except from the constant A, which is related to the energy

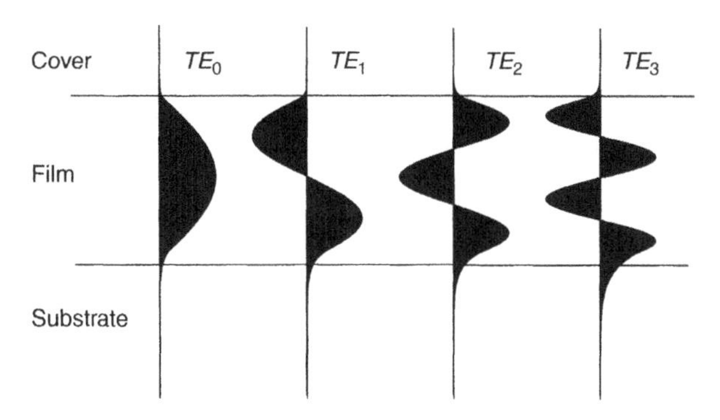

Figure 3.20 TE modes in an asymmetric step-index planar waveguide. The structure parameters are the following: $n_c = 1.00$, $n_f = 1.50$, $n_s = 1.43$, $d = 3.0$ μm, $\lambda = 633$ nm

carried by the mode. In addition, the integer m that characterises the mode order coincides with the number of zeros of the electric field profile function.

The electric fields in the cover and in the substrate are indeed evanescent waves, a particular case of inhomogeneous wave as discussed in the previous chapter, where the directions of propagation and attenuation are perpendicular. The modulus of the attenuation vector $\mathbf{a}_t$ defined in Section 2.1.6 is now given by γ_c in the cover region and by γ_s in the substrate. Therefore, the evanescent wave penetrations are determined by $1/\gamma_c$ and $1/\gamma_s$. As it can be observed in Figure 3.20, for a particular mode the field penetration in the cover is lower than in the substrate, and this is so because $n_c < n_s$, from (3.32) and (3.34) it follows that $\gamma_c > \gamma_s$. Another important feature related to evanescent fields is that as the mode number m increases, the wave penetration in a particular region is deeper (see the different modes in Figure 3.20). This behaviour is due to the fact that as the mode order increases, the propagation constant of the modes decreases, thus lowering the value of γ_c, which implies an increase in the field penetration.

Once the electric field component E_y of a particular guided mode has been established (the only non-vanishing electric field component for TE modes), the determination of the magnetic field associated with the mode is straightforwardly obtained by using the equations (3.11) and (3.12), which relate the H_x and H_z components of the magnetic field to the E_y component. In this way, the waveguide mode is fully characterised, leaving only the parameter A to be determined from the energy carried by the mode.

Guided TM modes In this case we are interested in the determination of the electromagnetic field structure within the planar waveguide based on the magnetic field, because in TM polarisation the magnetic field has a single component (H_y). The wave equation established for TM propagation in a homogeneous region was:

$$\frac{d^2 H_y}{dx^2} + \left[k_0^2 n^2 - \beta^2\right] H_y = 0 \tag{3.43}$$

Following a similar procedure to that performed for TE modes in the last paragraph, one obtains a transcendental equation for confined TM waveguide modes in terms of the normalised parameters:

$$\tan\left[V\sqrt{1-b}\right] = \frac{\dfrac{1}{\gamma_1}\sqrt{\dfrac{b}{1-b}} + \dfrac{1}{\gamma_2}\sqrt{\dfrac{b+a}{1-b}}}{1 - \dfrac{1}{\gamma_1\gamma_2}\dfrac{\sqrt{b(b+a)}}{(1-b)}} \tag{3.44}$$

which is the dispersion relation for TM guided modes for asymmetric step-index planar waveguides. In this equation we have defined, for the sake of simplicity, the parameters $\gamma_1 \equiv (n_s/n_f)^2$ and $\gamma_2 \equiv (n_c/n_f)^2 = \gamma_1 - a(1-\gamma_1)$. The solution for the magnetic field associated with TM polarised modes remains as:

$$H_y(x) = \begin{cases} Ae^{-\gamma_c x} & x \geq 0 \\ A\left[\cos\kappa_f x - \dfrac{n_f^2}{n_c^2}\dfrac{\gamma_c}{\kappa_f}\sin\kappa_f x\right] & -d < x < 0 \\ A\left[\cos\kappa_f d + \dfrac{n_f^2}{n_c^2}\dfrac{\gamma_c}{\kappa_f}\sin\kappa_f d\right] e^{\gamma_s(x+d)} & x \leq -d \end{cases} \tag{3.45}$$

In a similar way seen for TE modes, the solution for guided TM modes has exponentially decreasing behaviour in the cover and substrate, and a sinusoidal solution in the film region. At variance to that found for TE modes, in TM polarised modes, there exists a discontinuity in the first derivative of the magnetic field component $H_y(x)$ at $x = 0$ and $x = -d$, coming from the fact of the continuity condition of $(1/n^2)dH_y/dx$ at the interfaces.

The electric field associated with TM modes can now be obtained from equations (3.18) and (3.19), having thus completely characterised the electromagnetic field pattern of the guided TM mode.

Cut-off An important aspect concerning waveguides is to know what should be the minimum film width necessary for the waveguide support of a specific mode of order m, at a given wavelength. In this situation, the effective refractive index of this particular mode N should be very close to the substrate refractive index n_s, as it is shown schematically in Figure 3.21. In this case, it yields:

$$N \approx n_s \Rightarrow \quad b = (N^2 - n_s^2)/(n_f^2 - n_s^2) \approx 0 \tag{3.46}$$

In this situation, the mode is said to be at *cut-off*. If the film width decreases, the effective refractive index decreases, and the mode is not longer a guided mode, but a substrate radiation mode, giving rise to a *leaky mode*.

The normalised film thickness V for TE and TM modes at the cut-off is given by:

$$V_C^{\mathrm{TE}} = \tan^{-1}(a^{1/2}) + m\pi \qquad \text{TE modes} \tag{3.47}$$

$$V_C^{\mathrm{TM}} = \tan^{-1}(a^{1/2}/\gamma_2) + m\pi \quad \text{TM modes} \tag{3.48}$$

where V_C^{TE} and V_C^{TM} express the film width of the core waveguide, relative to the wavelength, necessary to support the mth mode in TE and TM propagation, respectively. From these relations, two important conclusions can be deduced:

(i) As n_c must be lower than n_f, it follows that $\gamma_2 = (n_c/n_f)^2 < 1$, and consequently it holds that $V_C^{\mathrm{TM}} > V_C^{\mathrm{TE}}$. This inequality implies that if a waveguide supports a TM mode of mth order, the waveguide also supports a TE mode of the same order. The reciprocal situation does not apply in general.

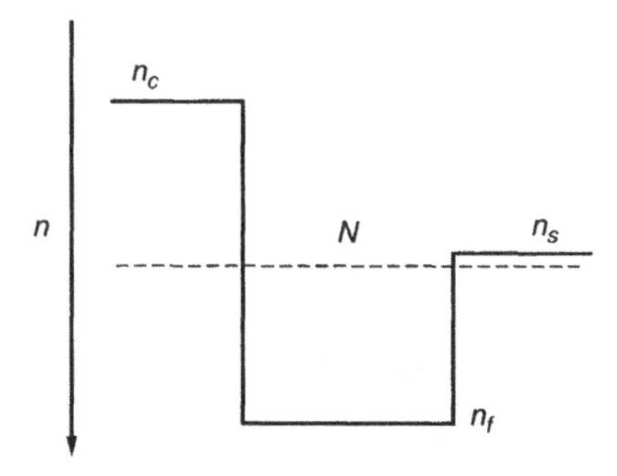

Figure 3.21 Position of the effective refractive index N, relative to the refractive indices of the waveguide structure, for a mode close to the cut-off

(ii) For a symmetric waveguide ($a = 0$), equations (3.41) and (3.44) yield $V_C^{TM} = V_C^{TE} = m\pi$. This indicates that a symmetric planar waveguide *always* supports at least the fundamental mode $m = 0$, both TE and TM polarised modes, regardless of the size (film thickness) or refractive indices of the guiding structure.

Radiation modes Up to now we have examined the solution of the wave equation for planar waveguides in terms of guided modes, where the radiation is mainly confined within the film, with decaying solution at both the cover and substrate regions in the form of evanescent waves. In this case, the mode effective index was restricted between the refractive index of the film and that of the substrate. Nevertheless, the wave equation, for both TE and TM polarisation light, also admits solutions for effective indices lower than n_s. In this case, we are dealing with *radiation modes*, where the light is no longer confined to the film, but can "leak" to adjacent regions, losing the light power inside the film core as the wave propagates along the waveguide. For this reason, these types of solutions are often called leaky modes.

Following the discussion of paragraph 3.3, outlined in Figure 3.16, for effective refractive index values lower than n_s but higher than n_c ($n_c < N < n_s$, $k_0 n_c < \beta < k_0 n_s$), the solutions in the film and substrate regions are in the form of oscillatory functions, while the behaviour of the fields in the cover region is in the form of exponential decay. This situation corresponds to *substrate radiation modes*, where the light is not confined to the film region, but also spreads out to the substrate, as can be seen in Figure 3.22. In addition, the solutions for leaky substrate modes are not discrete, but instead the wave equation for substrate modes admits an infinite number of solutions for continuous propagation constant values β (or effective refractive indices N).

Finally, if the mode effective refractive index N is lower than n_c ($N < n_c$, $\beta < k_0 n_c$) the solution for the modal fields in the three regions is in the form of sinusoidal functions. In this case the field pattern corresponds to a *radiation mode*, where the light cannot be confined in the film but leaks to the cover and substrate regions, as can be seen in Figure 3.23. Also, as in the case of substrate modes, there exists a continuous and infinite number of values for the propagation constant of radiation modes, with an infinite number of solutions for the electromagnetic field distribution.

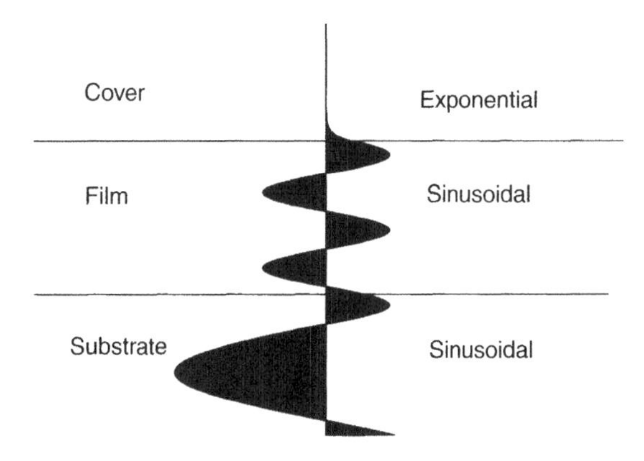

Figure 3.22 Substrate radiation mode in an asymmetric step-index planar waveguide

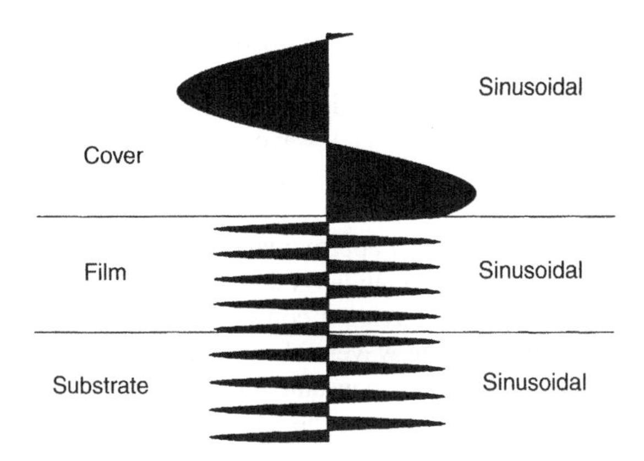

Figure 3.23 Radiation mode in an asymmetric step-index planar waveguide

3.5 Graded-index Planar Waveguides

The waveguide structures that we have examined up to now had a step-index profile, with three well-defined regions (cover, film and substrate) of constant refractive indices. These structures are obtained by any of the deposition techniques summarised in Chapter 1. When the waveguide fabrication is performed by a local index increase on a substrate material, by means of diffusion techniques, for instance, the refractive index of the waveguide shows in general a graded index profile in depth, expressed as $n = n(x)$. In this case the structure is called a *graded-index waveguide*. Usually the maximum index increase is located at the substrate surface, slowly decreasing in depth, until the refractive index reaches the substrate value, the typical depth being of the order of a few microns (Figure 3.24).

There are several methods that can be used to calculate the effective refractive index of the modes in waveguide structures with graded index profiles. Here we will describe two methods which are widely used and provide excellent results, namely, the multi-layer approximation and the ray approximation method. For both, we will discuss advantages and disadvantages, and the best choice for a given problem. We will also describe a method for reconstruction index profiles, given a discrete set of measured modal indices.

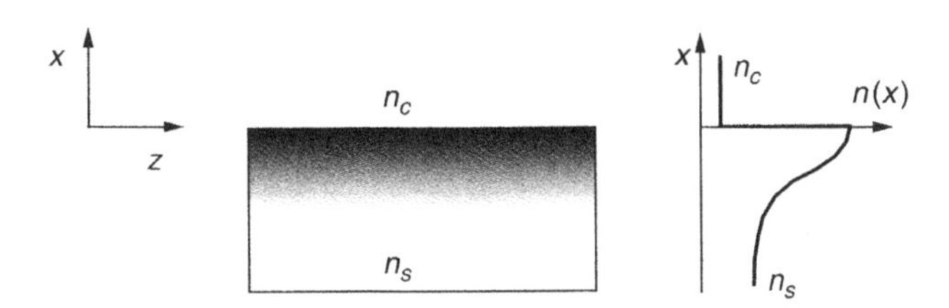

Figure 3.24 Graded-index planar waveguide, showing the typical graded index profile as a function of the depth

3.5.1 Multi-layer approximation

This method is based on the solutions obtained for asymmetric step-index planar waveguides, in such a way that the graded-index waveguide with an inhomogeneous core is decomposed into a finite number (as large as necessary) of homogeneous layers having constant refractive indices.

The operating method in multi-layer approximation is the following: first, the graded-index region, forming the core waveguide, is sectioned into p thin layers parallel to the planar interface, each of them having a constant refractive index n_j, as shown in Figure 3.25. In these conditions, the waveguide structure with graded refractive index $n(x)$ is defined by $(p+1)$ layers of constant refractive index n_j $(j = 0, 1, \ldots, p)$, where $n_0 = n_c$ and $n_p = n_s$. In addition, if the left boundary of the jth layer is situated at $x = x_j$, the individual layer thickness is automatically determined by $d_j = x_j - x_{j-1}$. Finally, the number of boundaries between two adjacent media is given by p.

If we restrict the solution of the graded waveguide to TE polarised modes, the non-vanishing electric field component will be expressed as:

$$\mathcal{E}_y = E_y(x)e^{i(\omega t - \beta z)} \tag{3.49}$$

The electric field amplitude $E_y(x)$ in each layer, which is assumed to be a homogeneous medium, should fulfil the wave equation for TE modes in planar waveguides:

$$\frac{d^2 E_y}{dx^2} + \left[k_0^2 n^2 - \beta^2\right] E_y = 0 \tag{3.50}$$

The general solution for the electric field amplitude E_y in a generic jth layer, as we have shown before, is given by:

$$E_j(x) = A_j e^{i\gamma_j(x-x_j)} + B_j e^{-i\gamma_j(x-x_j)} \quad (j = 0, 1, 2 \ldots, p) \tag{3.51}$$

where we have omitted the subscript y in the field amplitude for the sake of clarity. The x_j coordinate defines the left position of the jth layer, and the parameter γ_j is given by:

$$\gamma_j = \sqrt{k_0^2 n_j^2 - \beta^2} \quad (j = 0, 1, 2 \ldots, p) \tag{3.52}$$

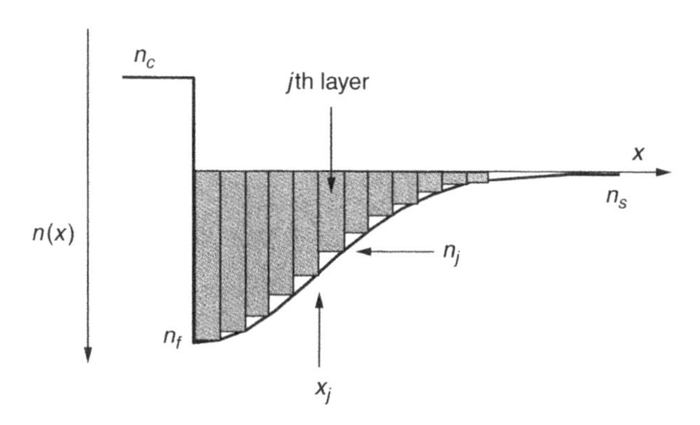

Figure 3.25 Sectioning of a graded-index planar waveguide for the multi-layer approximation analysis

being n_j the refractive index of the jth layer and β the propagation constant of the mode. The effective index of a mode is calculated, as usual, by $N = \beta/k_0$.

Considering the expressions given for the electric field amplitude E_j and the parameter γ_j given by (3.51) and (3.52) respectively, in the region where its refractive index n_j is higher than the effective refractive index of the mode N, the solution for the electric field is a sinusoidal function, while in a layer having a refractive index n_j lower than N the parameter γ_j is a pure imaginary quantity, and therefore the electric field will show exponential behaviour.

The A_j and B_j constants, which are in general complex magnitudes, will be determined by imposing the boundary conditions established for TE modes at the p interfaces, which are expressed as:

$$E_j = E_{j+1} \quad \text{at} \quad x = x_j \quad (j = 1, 2 \ldots, p) \tag{3.53}$$

$$dE_j/dx = dE_{j+1}/dx \quad \text{at} \quad x = x_j \quad (j = 1, 2 \ldots, p) \tag{3.54}$$

These two conditions allow us to express the A_{j+1} and B_{j+1} coefficients as functions of the A_j and B_j coefficients. The search for the propagation constant of a guided mode is carried out by establishing the condition of an exponentially decreasing behaviour of the electric field in the cover and substrate regions. Following this double condition, we start by imposing the boundary condition in the cover region, expressed as:

$$A_0 = 1 \quad \text{and} \quad B_0 = 0 \tag{3.55}$$

Next, we search for the propagation constants β that make the A_p coefficient vanish, that is, we impose the condition that in the substrate the behaviour of the field is an exponential decay. This value of the β parameter will correspond to a guided mode, having an effective refractive index of $N = \beta/k_0$.

The method described is applicable to any refractive index profile, and the solutions are as exact as needed, just by increasing the number of layers in which the graded index profile is divided. Moreover, once a particular propagation constant is calculated, the method gives the complete electric field profile of the guided mode along the transversal structure of the waveguide by using equation (3.51), and also the associated magnetic field using the appropriate formulae (equations (3.14) and (3.15)).

Another important advantage of the multi-layer approximation is the possibility of using complex refractive indices $(n_j + i\kappa_j)$ in any of the regions defining the structure. This issue is of particular interest when dealing with absorbing structures or with gain regions (in the case of lasers and amplifiers), or even structures including metallic layers such as in the case of control electrodes for integrated electro-optic modulators. In these cases, the propagation constant β becomes a complex quantity defined by $\beta = k_0(N_{eff} + iK_{\text{eff}})$, the imaginary part being directly related to the attenuation coefficient α (or gain coefficient if $\alpha < 0$) of the waveguide through the relation $K_{\text{eff}} = \alpha\lambda/4\pi$.

This approach has the inconvenience of using iterative methods to calculate the modal propagation constants. Additionally, the attenuation/gain calculation involves performing the iteration in the complex plane.

Figure 3.26 shows the refractive index profile corresponding to a graded-index planar waveguide, having a semi-Gaussian shape of the form:

$$n(x) = n_s + \Delta n e^{-(x/d)^2} \tag{3.56}$$

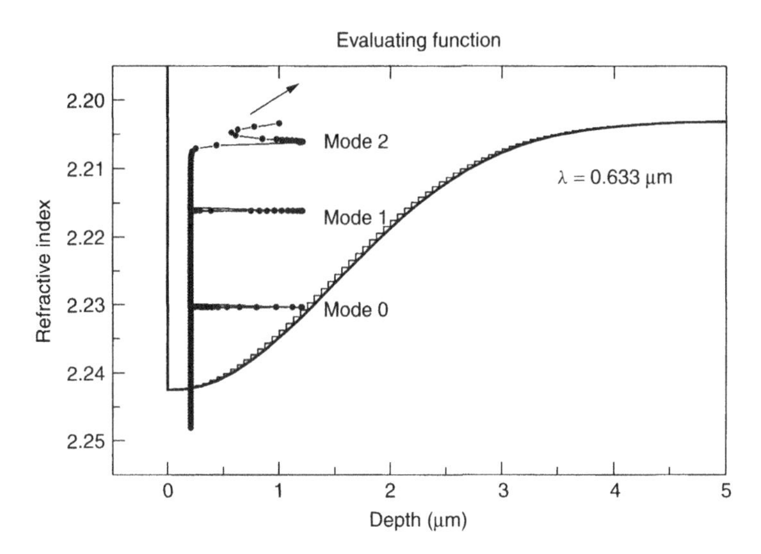

Figure 3.26 Semi-Gaussian refractive index profile of a graded index waveguide (thick line), besides the approximation by step-wise layers with constant refractive index (thin lines). Circles show the scan on effective refractive index to determine the position of the guided modes. Parameters of the waveguide and the modal positions are given in the text

with $n_s = 2.2030$, $\Delta n = 0.0395$ and $d = 2$ μm, being the maximum index at the surface $n_f = n_s + \Delta n$. The cover is assumed to be air ($n_c = 1.0000$). Besides the original graded index profile (thick line), we have drawn the sectioning of the profile by constant refractive index multi-layers (thin lines). In order to search for the propagation modes, we perform a discrete scan as a function of N (in the range n_f to n_s), calculating at each step the A_p coefficient. The guided modes correspond to effective refractive indices N for which the A_p coefficient vanishes. In Figure 3.26 we used the evaluation function $1/(1 + r^2)$, with $r \equiv |A_p|^2/|A_0|^2$, to determine the position of the minima on A_p (circles), using a wavelength of $\lambda = 0.633$ μm. These minima correspond to the maxima in the evaluation function, and define the modal positions supported by the graded-index waveguide. Following the graph drawn by the evaluation function, three maxima can be observed, that correspond to the guided modes at positions $N_0 = 2.2300$, $N_1 = 2.2157$ and $N_2 = 2.2059$. Clearly, these values are limited to the substrate refractive index (2.2030) and the maximum index at the surface (2.2425).

Once the modal positions have been determined, the field profile for each mode is calculated by using equation (3.51). Figure 3.27 shows the "mode levels" drawn as horizontal lines in the refractive index profile, and the associated electric field profile distributions. It is worth noting that the mode order coincides with the number of zeros of the field profile, as a general rule for modal distributions. Also, in a similar way to that pointed out in planar waveguides, the field penetration on the form of evanescent waves in the cover and in the substrate is deeper for higher order modes.

3.5.2 The ray approximation

The ray approximation method of calculating the modes in a graded-index planar waveguide is based on considering that a guided mode must fulfil the condition of

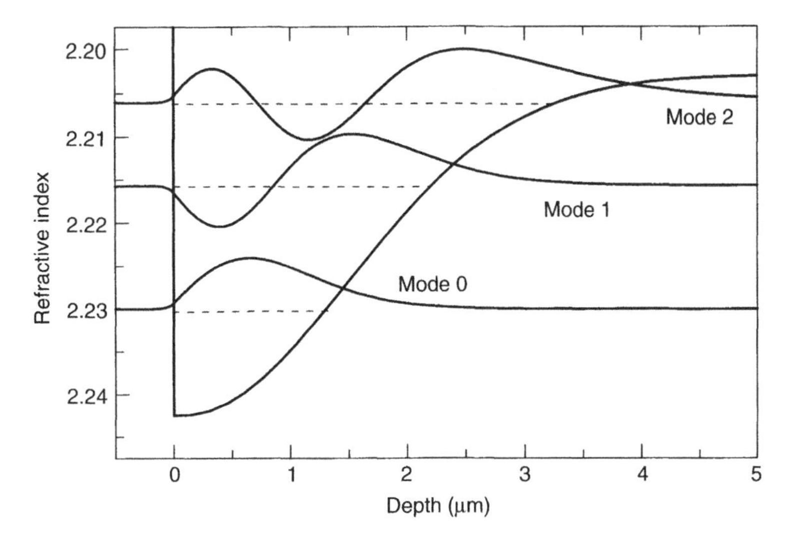

Figure 3.27 Position of the TE guided modes (dashed lines) and their corresponding electric field distributions, obtained by the multi-layer approximation method

constructive interference, such as that imposed to calculate the modes in an asymmetric planar waveguide by using the ray model. In the case of graded-index waveguides the problem arises from the fact that the total phase shift experienced by the ray must be calculated in each of the layers into which the graded waveguide must be decomposed [8].

The phase-shifts experimented by the wave propagating in a graded-index region can be determined by examining the path followed by a ray, as shown in Figure 3.28. As can be observed, the ray no longer follows a zig-zag path, but instead it changes its direction continuously, due to the fact that the refractive index changes in a continuous manner as a function of the depth. The propagation angle θ_j can easily be calculated by observing Figure 3.29:

$$\cos\theta_j = N/n(x_j) \tag{3.57}$$

where N is the effective refractive index of the guided mode, and $n(x_j)$ is the refractive index of the graded waveguide evaluated at a depth $x = x_j$. The relation between the position change in depth of the ray and its advance in the propagation direction is

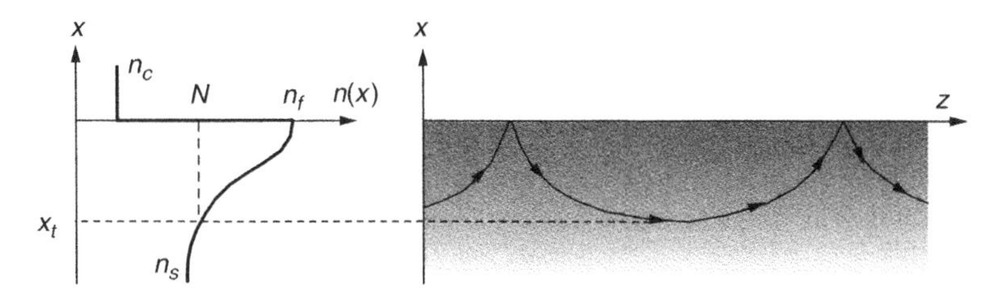

Figure 3.28 Left: graded refractive index profile, showing the position of the effective index position N of the mode and the corresponding turning point x_t. Right: curved path followed by a ray propagating within the graded index waveguide

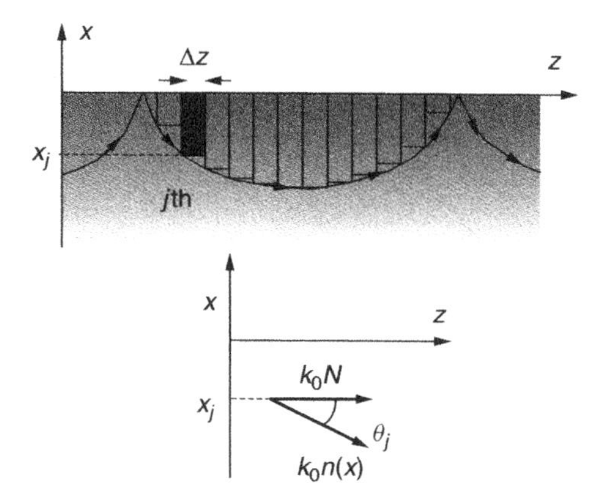

Figure 3.29 Above: Scheme of the partial displacements of the ray along the waveguide. Below: relation between the refractive index at the jth position, the effective index of the mode and the propagation angle

given by:

$$\Delta x_j = \Delta z \tan\theta_j \tag{3.58}$$

The maximum penetration of the ray in the graded index region is called the *turning point*, and can be determined by imposing the condition of $\theta_j = 0$ (see Figures 3.28 and 3.29). According to equation (3.57) this position is situated at a depth x_t that fulfils $n(x_t) = N$ (see Figure 3.28). Obviously, the turning point will be different for each guided mode, because of the different effective refractive indices associated to them. In addition, if the graded-index profile is monotonously decreasing in depth, higher order modes will have deeper turning point.

The incremental phase shift $\Delta\phi_j$ experienced by the ray after advancing a step distance in the depth direction is given by:

$$\Delta\phi_j = k_0 n(x_j)\sin\theta_j \Delta x_j = k_0\sqrt{n^2(x_j) - N^2}\Delta x_j \tag{3.59}$$

If we further assume that the index increase is small ($\Delta n \ll n_s$), the phase shifts from equations (2.148) and (2.149) in the turning points are given by:

$$\phi_0 = \pi \qquad (\text{at } x = 0) \tag{3.60}$$

$$\phi_t = \pi/2 \quad (\text{at } x = x_t) \tag{3.61}$$

The condition for constructive interference is expressed imposing an integral multiple of 2π on the phase shift after a complete round trip of the ray. Thus, the mathematical condition that should fulfil a guided mode yields:

$$\Sigma\Delta\phi_j - \phi_c - \phi_t = 2m\pi \tag{3.62}$$

By using the partial contributions to the phase shift from equation (3.59), and taking into account the phase shifts at the surface boundary and at the turning point (equations (3.60) and (3.61)), the last condition for the guided modes is finally given by:

$$2k_0 \int_0^{x_t} \sqrt{n^2(x) - N^2 dx} = \left(2m + \frac{3}{2}\right)\pi \tag{3.63}$$

In general, the graded-index profiles generated by diffusion processes for waveguide fabrication can be expressed in the following general form:

$$n(x) = n_s + \Delta n f(x/d) \tag{3.64}$$

where n_s is the substrate refractive index and Δn is the maximum index change, usually located at the surface. The function $f(x/d)$ determines the particular form (exponential, Gaussian, error function, etc.) of the index increase as function of the depth where the parameter d indicates the *diffusion* or *penetration depth*. Also, it is further assumed that the function f verifies $0 < f(x/d) < 1$.

As it was done for the case of asymmetric step-index planar waveguides, it is worth defining a set of normalised parameters by:

$$b \equiv \frac{N^2 - n_s^2}{n_f^2 - n_s^2} \quad \text{Normalised mode index} \tag{3.65}$$

$$V_d \equiv k_0 d\sqrt{n_f^2 - n_s^2} \quad \text{Normalised penetration depth} \tag{3.66}$$

where n_f stands for the maximum index change at the surface, given by $n_f = n_s + \Delta n$. In most of the cases, the diffusion process performed to fabricate waveguides gives rise to small refractive index increases relative to the substrate, and under this circumstance it holds that $\Delta n \ll n_s$, and thus the following approximation can be made:

$$n^2(x) \approx n_s^2 + (n_f^2 - n_s^2) f(x/d) \tag{3.67}$$

Using this relation, the equation for calculating the guided modes can be expressed as:

$$2V_d \int_0^{\xi_t} \sqrt{f(\xi) - b d\xi} = \left(2m + \frac{3}{2}\right)\pi \tag{3.68}$$

where $\xi \equiv x/d$, $\xi_t \equiv x_t/d$ and $b \equiv f(\xi_t)$. In this way, for a particular given function $f(\xi)$, containing the information of the refractive index profile shape, the last equation can be numerically solved.

In the particular case of a Gaussian function $f(\xi) = exp(-\xi^2)$, which is the most common profile obtained in waveguide fabrication by diffusion processes, the cut for the mth mode is given by the compact formula:

$$V_{\text{dm}} = (2\pi)^{1/2}(m + 3/4) \tag{3.69}$$

where we have used facts that in the cut-off condition it holds that $b \approx 0$ and $x_t \to \infty$, and the result $\int_0^\infty \sqrt{e^{-\xi^2}} d\xi = \sqrt{\pi/2}$.

On the other hand, if the refractive index distribution of the graded-index waveguide is symmetric ($f(\xi) = f(-\xi)$), the next set of equations is obtained:

$$2\Sigma\phi_j - 2\phi_t = 2m\pi \quad \text{Resonant condition for guided modes} \tag{3.70}$$

$$2V_d \int_0^{\xi_t} \sqrt{f(\xi) - b d\xi} = \left(m + \frac{1}{2}\right)\pi \quad \text{Condition for guided modes} \tag{3.71}$$

$$V_{\text{dm}} = (\pi/2)^{1/2}(m + 1/2) \quad \text{Cut-off condition} \tag{3.72}$$

The last equations are valid for TE modes as well as for TM modes, providing that the condition $\Delta n \ll n_s$ holds. Indeed, this relation is widely fulfilled in practice in most cases.

The advantage of this method stems from the easy numerical implementation, and also that the method is valid for general smooth index profiles. The most important drawback of the ray model method lies in the fact that, although the propagation constants of the guided modes are accurately obtained, it does not provide information about the distribution of the modal fields.

3.5.3 Reconstruction of index profiles: the inverse WKB method

One of the most important issues in waveguide technology is to reconstruct the index profile of a planar waveguide from a knowledge of the propagation constant of their guided modes. Construction of the refractive index profile from the effective indices is not only important for waveguide prescriptions but also for the evaluation of existing waveguides. Given a refractive index profile $n(x)$, the guided-mode equation uniquely defines an effective index set of guided modes $N(m)$, m being an integer number, but conversely, an infinite number of profiles that provide the same set of effective indices can be found. Therefore, in some cases, it is necessary to restrict the solutions to physically reliable index profiles.

The characteristic equation for the guided mth-order mode given in (3.62) can be expressed as:

$$2k_0 \int_0^{x_t(m)} \sqrt{n^2(x) - N^2(m)}dx - \phi_c - \phi_t = 2m\pi \tag{3.73}$$

being the upper integral limit related to the refractive indices by:

$$n(x_t(m)) = N(m) \tag{3.74}$$

where $n(x)$ is the refractive index profile that defines the waveguide, $N(m)$ is the effective refractive index of the mth-order guided mode, and $x_t(m)$ is the turning point of the mth mode, defined as the depth at which the refractive index is equal to the effective index (Figure 3.28). The quantities ϕ_c and ϕ_t denote the phase changes at the surface $x = 0$ and at the turning point x_t beneath the surface. If the variation of the refractive index at the turning point is sufficiently slow, it can be demonstrated [9] that the phase change ϕ_t is given by $\pi/2$.

One way of reconstructing the index profile is to invert the modal equation (3.73) by computing the locations of turning points x_t recursively [10]. This equation is formally identical to that obtained by applying the WKB approximation used in quantum mechanics to solve the one-dimensional Shrödinger equation, thus the method for reconstructing the refractive index profile starting from equation (3.73) is known as the *inverse WKB method* (IWKB). Nevertheless, the direct inversion of equation (3.73) is accurate only for highly multi-mode waveguides, since the number of straight line segments is equal to the number of guided modes.

An elegant way of avoiding this inconvenience is to define a continuous *effective-index function*, which is then used to construct a refractive index profile by numerically solving the WKB equation [11]. This procedure involves two steps: (a) to construct an effective index function $N(q)$ from the discrete set of measured guided modes, and (b) to determine the corresponding profile $n(x)$ by inverting equation (3.73).

Having a set of ν experimentally measured modal indices $N(m)$ ($m = 0, 1, 2 \ldots, \nu - 1$), a natural way to construct the artificial effective index function $N(q)$ is to fit the ν discrete modal positions to a polynomial of order $(\nu - 1)$. This can be done for instance by Neville's algorithm [12]. As an example of this implementation, Figure 3.30 shows four measured mode indices (circles), and the associated effective index function calculated by a polynomial fit of order 3 (continuous line).

In order to obtain the peak refractive index n_0 at the surface we first need to estimate the value $q = q_0$, which is done by substituting $x_t = 0$ into equation (3.73), where the guided mode indices m are now denoted by the artificial set of modes of indices q. In this case, by substituting $n_0 = n(0) = N(q)$ $(= n_f)$ in equation (2.149), we obtain $\phi_s = \pi$. Substituting this value into equation (3.73), besides $\phi_t = \pi/2$, and taking into account that the WKB integral vanishes, we finally obtain that the value of q at the surface is $q_0 = -0.75$. Therefore, the refractive index at the surface is easily computed by evaluating the index function, obtained by a polynomial fit, at $q = -0.75$:

$$n_0 = n(0) = N(q_0) = N(-0.75) \tag{3.75}$$

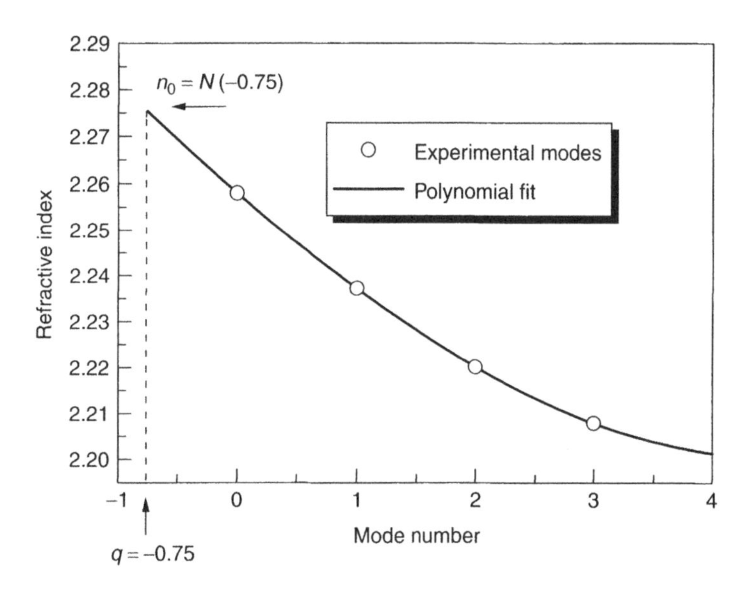

Figure 3.30 Effective index function $N(q)$ (continuous line) determined by a polynomial fit from the experimental values of modal refractive indices $N(m)$ (circles). The surface peak index of the refractive index distribution n_0 is calculated by evaluating the index function at $q = -0.75$ (arrows)

The polynomial fit shown in Figure 3.30 gives a refractive index value at the surface of $n_0 = 2.2751$.

Once the refractive index at $x = 0$ has been evaluated, the construction of the index profile $n(x)$ is point-wise built by a recursive algorithm. First, we sample $N(q)$ in descending order $N_0 > N_1 > N_2 > \ldots$, not necessarily equally spaced, where N_i corresponds to a turning point x_i. Therefore, all we need to reconstruct the index profile is to compute x_i locations. To determine x_i we use step-wise approximation to the profile, and the index N_i of the ith step is given by the average value of N_i and N_{i-1}, for $i = 1, 2, \ldots$. This is valid provided that x_i and x_{i+1} are sufficiently close to each other.

Equation (3.73) can be rewritten, separated into two terms as:

$$I_i = \alpha_i \tag{3.76}$$

where we have defined:

$$I_i = k_0 \int_0^{x_i} \sqrt{n^2 - N_i^2}\, dx \tag{3.77}$$

and:

$$\alpha_i = q_i\pi + \phi_c(N(q_i))/2 + \pi/4 \tag{3.78}$$

The relations (3.76)–(3.78) stand for $i = 1, 2, 3 \ldots$. On the other hand, the integral given by (3.77) can be replaced by a finite sum of the form:

$$I_i = k_0 \left\{ x_i \sqrt{\overline{N}_i^2 - N_i^2} + \sum_{j=1}^{i-1} x_j \left[\sqrt{\overline{N}_j^2 - N_i^2} - \sqrt{\overline{N}_{j+1}^2 - N_i^2} \right] \right\} \tag{3.79}$$

Equation (3.76) provides a way to design an algorithm to compute x_i's in a recursive manner. By rearranging (3.79) and using equation (3.76) we obtain:

$$x_i = \frac{\alpha_i - \sum_{j=1}^{i-1} \left\{ x_j \left[\sqrt{\overline{N}_j^2 - N_i^2} - \sqrt{\overline{N}_{j+1}^2 - N_i^2} \right] \right\}}{k\sqrt{\overline{N}_i^2 - N_i^2}} \tag{3.80}$$

for $i = 1, 2, 3, \ldots$ and $x_0 = 0$. Thus, starting from the fitted index function $N(q)$, equation (3.80), as well as equation (3.78), gives a straightforward way of reconstructing the refractive index profile $n(x)$.

Figure 3.31 shows an example of refractive index profile reconstruction using the IWKB algorithm described. The procedure starts from a discrete set of measured indices $N(m)$, which we fit into a polynomial function, to obtain the effective index function $N(q)$ (Figure 3.30). Then, we use the recursive algorithm given by expression (3.80) to obtain the turning points x_i's associated with each effective index. Figure 3.31 shows the refractive index profile calculated by the IWKB method (circles), besides the exact profile (continuous line), where the working wavelength is $\lambda = 0.633$ μm. As can be seen, the IWKB profile reproduces the real profile very accurately, in spite of the low number of modes (four) used to calculate the index profile. It is worth noting that the

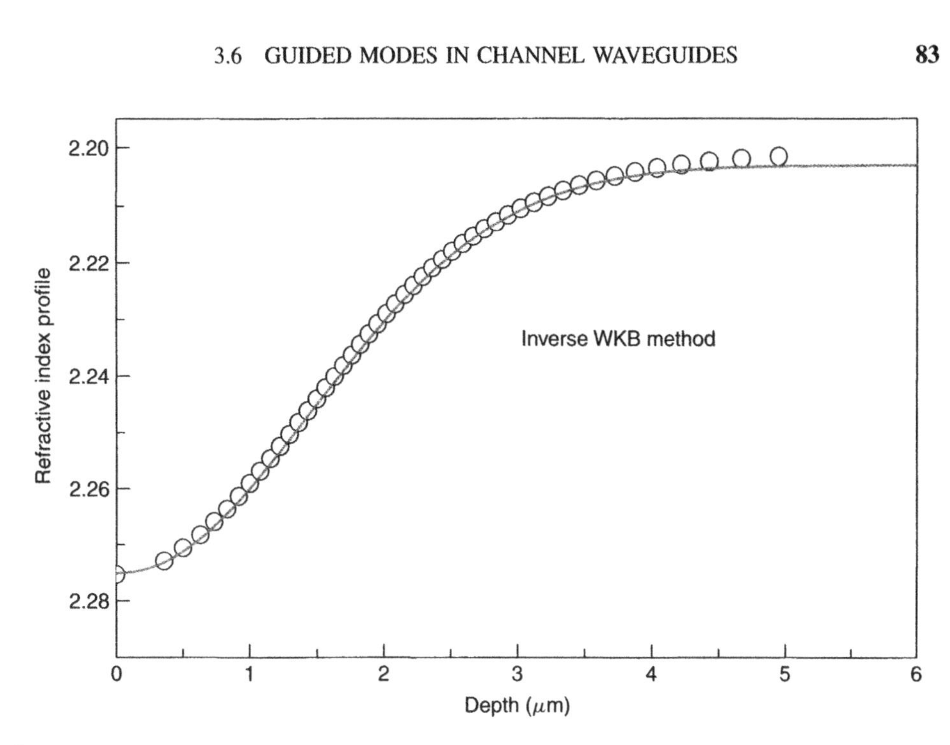

Figure 3.31 Reconstruction of the refractive index profile by the IWKB method (circles) using the data given in Figure 3.30. Continuous line is the exact index profile

refractive index peak is quite well estimated, and the fit is only less accurate for very deep regions, where the refractive index of the waveguide reaches that of the substrate.

Due to the inevitable existence of errors in measured effective indices, in practice it is convenient to employ least-squares fitting to construct the effective index function for the reconstruction of refractive index profiles. In this way, the influence of experimental errors in the obtained index profile can be minimised.

3.6 Guided Modes in Channel Waveguides

In general, the integrated photonic devices presented in Chapter 1 are based on channel (or 2D) waveguides, in which the light is confined in two directions, allowing the propagation in only one direction, in contrast with the planar waveguides studied in the previous sections, where light is confined only in the direction perpendicular to the interfaces. In this way, radiation travelling in channel waveguides can propagate without suffering diffraction, that will otherwise give rise to power loss. Therefore, for performing functions such as modulation, switching, amplification, etc., the channel waveguide is the right choice for the fabrication of integrated optical devices.

The most common geometries used for the definition of channel waveguides in integrated photonic devices are the stripe and the buried waveguides (Figure 3.32). The rib waveguide can be considered to be special case of stripe waveguide, as was explained in Section 3.1. Stripe optical waveguides are widely used for semiconductor-based photonic chips, such as GaAs or InP, and also in polymeric-based integrated photonic devices. Channel waveguide fabrication involves a selective etching of a high index film previously deposited onto a low index substrate. The etching can be

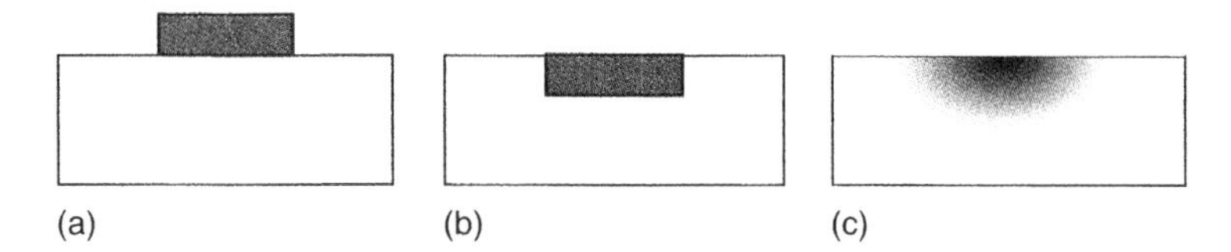

Figure 3.32 Channel or 2D waveguides: (a) load type or stripe channel waveguide; (b) buried channel waveguide; (c) graded-index buried channel waveguide

performed by means of physical methods (ion milling) or chemical methods (solvents, acids, etc.), or even by a combination of both, such as in reactive ion etching (RIE) [13]. In general, stripe and rib waveguides tend to have relatively high propagation losses (~1 dB/cm) due to the roughness of the top and lateral walls which define the optical channels. One way of reducing losses in these waveguides is to deposit a cladding material covering the channels, which also serves as a protection layer against environmental chemical agents.

Buried waveguides are fabricated by the refractive index increase of a substrate, in regions previously defined by appropriate photolithographic masks. The index increase is usually carried out by diffusion processes, and because of that, the channel waveguides fabricated following this method give rise to graded-index profiles [14]. The main advantage of this type of channel waveguides, typical in glasses and ferro-electric materials, is the low propagation losses that can be achieved (less than 0.1 dB/cm). Also, buried channel waveguide geometry allows easy placing of the metallic control electrodes, such as in the case of electro-optic modulators and switches.

When dealing with planar waveguides, whether step-index or graded index structures, light propagation can be described in terms of two mutually orthogonal polarisations, namely, the TE and TM propagating modes. In contrast, in channel optical waveguides there are no pure TE or TM modes, but instead there are two families of hybrid transversal electromagnetic modes (TEM). Fortunately, the TEM modes that propagate in channel waveguides are strongly polarised along the x or y direction (z being the direction of propagation of light), and therefore a classification can be made according to the major component of the electric field associated with the electromagnetic radiation. Optical modes having the main electric field component along the x axis are called E^x_{pq} modes, and behave very similarly to the TM modes in a planar waveguide. For this reason, they are known as *quasi-TM modes*. The subscripts p and q denote the number of nodes of the electric field E_x in the x and y direction, respectively. Accordingly, the E^y_{pq} modes have E_y as the major component of the electric field, and are closely related to the TE modes in a planar waveguide, and can be considered as *quasi-TE modes*.

An exact treatment of the modal characterisation in 2D waveguides is not possible, even in the simplest case of a symmetric rectangular channel waveguide. Therefore, in order to solve this problem, some approximation should be made, and there are several numerical methods which yield good results in general. Here we will explain two widely used methods: Marcatili's method and the effective index method. While the first one allows us to calculate the electromagnetic field in a rectangular channel waveguide (with a homogeneous central core), with the latter we can obtain the optical modes supported by a channel waveguide with arbitrary geometry, even with graded index regions (whether the core or the surroundings).

3.6.1 Marcatili's method

This approximate method can be used to calculate the propagation constants and modal fields supported by a rectangular channel waveguide, whether stripe or buried, as the one shown in Figure 3.33.

Marcatili's method [15] allows us to model a channel waveguide geometry as shown in Figure 3.34, which consists of a central homogeneous high index core surrounded by four homogeneous low index regions. The waveguide core, referred to as region I in the figure, has a rectangular cross-section with dimensions a and b in the x and y directions respectively, and a refractive index n_1. The central core is surrounded by homogeneous regions II, III, IV and V as indicated in Figure 3.34, which have refractive indices n_2, n_3, n_4 and n_5 respectively.

If the propagation constant β of the mode is far from the cut-off ($\beta \approx k_0 n_1$), the electromagnetic field is confined mainly in the core (region I), and only a small fraction of the energy carried by the optical mode spreads out to the surrounding regions (regions II, III, IV and V). Moreover, the fields penetrate even less in the four corners

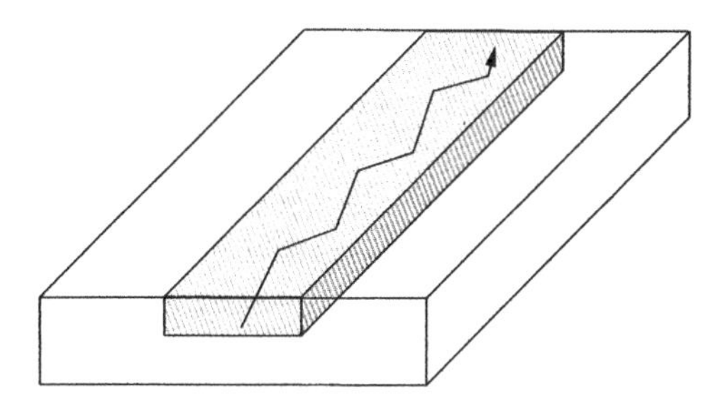

Figure 3.33 Geometry of a rectangular buried channel waveguide, which can be modelled by Marcatili's method

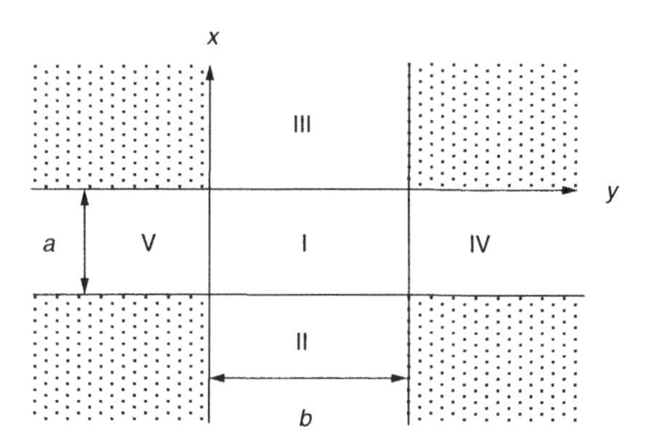

Figure 3.34 Channel waveguide geometry used for modal analysis employing Marcatili's method. The dotted regions are not considered in this analysis

(dotted regions in Figure 3.34), and therefore in these regions there is very little energy of the mode. This is the argument used in Marcatili's method to completely ignore these regions, and thus the analysis can be greatly simplified.

As shown in Section 3.3, if the refractive index of a medium depends only on the x and y coordinates, and we choose the z direction as the propagation direction, the electric field for harmonic waves can be expressed in a general form as:

$$\mathcal{E}(x, y, z, t) = \mathbf{E}(x, y)e^{i(\omega t - \beta z)} \tag{3.81}$$

$$\mathcal{H}(x, y, z, t) = \mathbf{H}(x, y)e^{i(\omega t - \beta z)} \tag{3.82}$$

By substituting the proposed solution given by (3.81) and (3.82) into Maxwell's equations (3.6) and (3.7), in a similar way as we proceeded to obtain the wave equations for planar waveguides, this allows us to express the transversal components of the electric and magnetic fields as functions of their longitudinal components, yielding:

$$E_x = -(i/K_j^2)[\beta(\partial E_z/\partial x) + \omega\mu_0(\partial H_z/\partial y)] \tag{3.83}$$

$$E_y = -(i/K_j^2)[\beta(\partial E_z/\partial y) - \omega\mu_0(\partial H_z/\partial x)] \tag{3.84}$$

$$H_x = -(i/K_j^2)[\beta(\partial H_z/\partial x) - \omega n_j^2\varepsilon_0(\partial E_z/\partial y)] \tag{3.85}$$

$$H_y = -(i/K_j^2)[\beta(\partial H_z/\partial y) + \omega n_j^2\varepsilon_0(\partial E_z/\partial x)] \tag{3.86}$$

where we have introduced the parameter K_j defined as:

$$K_j^2 = n_j^2 k_0^2 - \beta^2 \tag{3.87}$$

and k_0 is given as usual by $k_0 = \omega^2\varepsilon_0\mu_0$. Taking into account the last four equations, a complete solution for the electromagnetic optical fields can be established once the solutions for their longitudinal components E_z and H_z are known.

Besides the relations (3.83)–(3.86), Maxwell's equations allow us also to establish a equation for the longitudinal component of the electric field E_z and a similar equation for the longitudinal component of the magnetic field H_z. These two equations are referred to as the reduced wave equations for channel waveguides, and are given by:

$$\frac{\partial^2 E_z}{\partial x^2} + \frac{\partial^2 E_z}{\partial y^2} = [\beta^2 - k_0^2 n^2(x, y)]E_z \tag{3.88}$$

$$\frac{\partial^2 H_z}{\partial x^2} + \frac{\partial^2 H_z}{\partial y^2} = [\beta^2 - k_0^2 n^2(x, y)]H_z \tag{3.89}$$

Then, the calculation of propagating modes in channel waveguides involves the resolution of the wave equations (3.88) and (3.89), taking into account the appropriate boundary conditions that must fulfil the fields at the interfaces.

E_{pq}^x *polarised modes* Marcatili's method for calculating the quasi-TM polarised modes in a rectangular channel waveguide, where the electric field is polarised mainly along the x direction, is based on the separation of variables, in such a way that the components of the electric and magnetic fields can be factorised as a product of two functions, one of them having dependence only on the x coordinate and the second one which depends only on the y coordinate.

Region I As we are interested in confined modes, the propagation constant β must be lower than $n_1 k_0$, being n_1 the refractive of the core. Therefore the longitudinal components of the fields for quasi-TM modes in region I should be expressed in terms of sinusoidal functions of the form:

$$E_z = A\cos\kappa_x(x+\xi)\cos\kappa_y(y+\eta) \tag{3.90}$$

$$H_z = -A(\varepsilon_0/\mu_0)^{1/2} n_1^2(\kappa_y/\kappa_x)(k_0/\beta)\sin\kappa_x(x+\xi)\sin\kappa_y(y+\eta) \tag{3.91}$$

As the longitudinal components E_z and H_z must fulfil the wave equations (3.88) and (3.89) respectively, substituting E_z and H_z in the wave equations a relation between the parameters κ_x and κ_y and the propagation constant β can be found:

$$K_1^2 = n_1^2 k_0^2 - \beta^2 = \kappa_x^2 + \kappa_y^2 \tag{3.92}$$

Besides this relation, the coefficients κ_x and κ_y are still unknown. These two coefficients, in addition to the phase parameters ξ and η (and others that will be necessary to introduce for the fields description in the remaining regions), are determined by imposing the appropriate boundary conditions that must fulfil the electric field and magnetic field at the four boundaries.

The transversal components of the electric and magnetic fields are obtained by substituting the proposed solutions given in (3.90) and (3.91) into the equations (3.83)–(3.86), yielding:

$$E_x = (iA/\kappa_x\beta)(n_1^2 k_0^2 - \kappa_x^2)\sin\kappa_x(x+\xi)\cos\kappa_y(y+\eta) \tag{3.93}$$

$$E_y = -iA(\kappa_y/\beta)\cos\kappa_x(x+\xi)\sin\kappa_y(y+\eta) \tag{3.94}$$

$$H_x = 0 \tag{3.95}$$

$$H_y = iA(\varepsilon_0/\mu_0)^{1/2} n_1^2(k_0/\kappa_x)\sin\kappa_x(x+\xi)\cos\kappa_y(y+\xi) \tag{3.96}$$

For well-confined modes, having most of their energy concentrated in the core, it holds that their propagation constants are far from cut-off ($\beta \approx n_1 k_0$), and taking into account the relation (3.92), this implies that:

$$\kappa_x, \kappa_y \ll \beta \tag{3.97}$$

and thus from equation (3.94) it follows that $E_y \ll E_z$, and we can ignore the component E_y as a magnitude of second order in κ_y/β. As the H_x component is zero, our analysis is based on considering only four non-vanishing components of the fields, in particular E_x, E_z, H_z and H_y. In addition, as we are dealing with propagating modes polarised mainly along the x direction, we find that $E_z \ll E_x$.

The fields in the adjacent regions to the core (regions II, III, IV and V) must disappear for long distances, implying that the solutions for the fields in these regions should be in the form of exponential decays. An additional condition that must fulfil the fields is that the tangential component of the electric field must be continuous at each boundary. Considering all these requirements, the following solutions for the longitudinal components of the electric and magnetic fields in the regions II, III, IV and V can be obtained.

Region II By imposing the condition of continuity for the E_z component at $x = -a$, and considering an exponential decay behaviour in the x direction, the longitudinal component of the fields in region II is given by:

$$E_z = A\cos\kappa_x(\xi - a)\cos\kappa_y(y+\eta)\,exp[\gamma_2(x+a)] \tag{3.98}$$

$$H_z = -A(\varepsilon_0/\mu_0)^{1/2}n_2^2(\kappa_y/\gamma_2)(k_0/\beta)\cos\kappa_x(\xi - a)\sin\kappa_y(y+\eta)\,exp[\gamma_2(x+a)] \tag{3.99}$$

where:

$$K_2^2 = n_2^2k_0^2 - \beta^2 = \kappa_y^2 - \gamma_2^2 \tag{3.100}$$

Region III In this case the condition of continuity must be imposed on the E_z component at the position $x = 0$. After a similar procedure done to calculate the fields in region II, we obtain:

$$E_z = A\cos\kappa_x\xi\cos\kappa_y(y+\eta)\,exp(-\gamma_3 x) \tag{3.101}$$

$$H_z = A(\varepsilon_0/\mu_0)^{1/2}n_3^2(\kappa_y/\gamma_3)(k_0/\beta)\cos\kappa_x\xi\sin\kappa_y(y+\eta)\,exp(-\gamma_3 x) \tag{3.102}$$

where:

$$K_3^2 = n_3^2k_0^2 - \beta^2 = \kappa_y^2 - \gamma_3^2 \tag{3.103}$$

Region IV In this case the amplitudes must be adjusted in order that the majority component of the electric field E_x is continuous across the boundary at the position $y = b$, obtaining:

$$E_z = A(n_1^2/n_4^2)\cos\kappa_y(b+\eta)\cos\kappa_x(x+\xi)\,exp[-\gamma_4(y-b)] \tag{3.104}$$

$$H_z = -A(\varepsilon_0/\mu_0)^{1/2}n_1^2(\gamma_4/\kappa_x)(k_0/\beta)\cos\kappa_y(b+\eta)\sin\kappa_x(x+\xi)\,exp[-\gamma_4(y-b)] \tag{3.105}$$

where:

$$K_4^2 = n_4^2k_0^2 - \beta^2 = \kappa_x^2 - \gamma_4^2 \tag{3.106}$$

Region V In the same way as the procedure following for calculating the fields in region IV, we impose the continuity of E_x at the boundary between regions I and V ($y = 0$), and the result is:

$$E_z = A(n_1^2/n_5^2)\cos\kappa_y\eta\cos\kappa_x(x+\xi)\,exp(\gamma_5 y) \tag{3.107}$$

$$H_z = A(\varepsilon_0/\mu_0)^{1/2}n_1^2(\gamma_5/\kappa_x)(k_0/\beta)\cos\kappa_y\eta\sin\kappa_x(x+\xi)\,exp(\gamma_5 y) \tag{3.108}$$

where:

$$K_5^2 = n_5^2k_0^2 - \beta^2 = \kappa_x^2 - \gamma_5^2 \tag{3.109}$$

In addition to the condition of continuity of the tangential component of the electric fields at the boundaries, it is necessary to impose also the boundary condition of the continuity of the magnetic field component H_z at the interfaces $x = 0$ and $x = -a$. From these two conditions we obtain the following transcendental equations:

$$\tan(\kappa_x a) = n_1^2\kappa_x(n_3^2\gamma_2 + n_2^2\gamma_3)/(n_3^2n_2^2\kappa_x^2 - n_1^4\gamma_2\gamma_3) \tag{3.110}$$

$$\tan(\kappa_x\xi) = -(n_3^2/n_1^2)(\kappa_x/\gamma_3) \tag{3.111}$$

From equation (3.110) the coefficient κ_x can be numerically calculated, because the coefficients γ_2 and γ_3 can be expressed as function of κ_x using (3.92) as well as the relations (3.100) and (3.103). The second transcendental equation (3.111) allows us to calculate the phase parameter ξ.

In order to determine the propagation constant β using the expression (3.92) it is also necessary to calculate the value of the parameter κ_y. This parameter can be determined after imposing the continuity of the longitudinal component of the magnetic field H_z at the boundaries between region I and regions IV and V, that is, at the positions $y = b$ and $y = 0$. These two conditions generate two new transcendental equations given by:

$$\tan(\kappa_y b) = \kappa_y(\gamma_4 + \gamma_5)/(\kappa_y^2 - \gamma_4\gamma_5) \tag{3.112}$$

$$\tan(\kappa_y \eta) = -\gamma_5/\kappa_y \tag{3.113}$$

From the transcendental equation (3.112), as well as the equations (3.92), (3.106) and (3.109), the parameter κ_y can be calculated numerically. Once κ_y has been determined, as well as the value of κ_x previously calculated, the propagation constant of the mode can be obtained by means of the expression:

$$\beta = [n_1^2 k_0^2 - (\kappa_x^2 + \kappa_y^2)]^{1/2} \tag{3.114}$$

Finally, the only remaining parameter to be determined is the phase parameter η, which is obtained from the transcendental equation (3.113). In this way, once all the parameters have been determined, the electromagnetic field distribution of the mode at the five regions of the rectangular channel waveguide is completely established. Indeed, once E_z and H_z are known, the remaining components of the fields can easily be obtained from equations (3.83)–(3.86).

Because the transcendental relations involve the tangent function, in general we obtain a discrete number of solutions for the parameters κ_x and κ_y. Therefore, the propagation constant β associated with a mode is determined by two integer numbers p and q ($p = 0, 1, 2 \ldots$, $q = 0, 1, 2, \ldots$), and the quasi-TM mode is denoted by E_{pq}^x.

Figure 3.35 shows the modal intensity distributions of the quasi-TM modes (E_{pq}^x polarised modes) at $\lambda = 1.3$ μm obtained by Marcatili's method. The channel waveguide is a rectangular buried waveguide 2 μm deep and 5 μm wide, having a core refractive index (region I) of 1.55. Regions II, IV and V correspond to the substrate medium, which has a refractive index of 1.45. The upper part of the channel waveguide is air, and therefore region III has a refractive index of $n_3 = 1.00$.

From Figure 3.35 we observe that the TM_{00} mode has most of its energy concentrated in region I, and therefore the initial assumption made for the Marcatili approximation is well fulfilled. For higher order modes, although the energy is still concentrated in region I, the mode penetrates appreciably in the adjacent regions (II, III, IV and V); the higher the mode order, the deeper the penetration into the adjacent media. Also, in region III there is very little intensity due to the large difference between its refractive index and the refractive index of the core region ($n_3 = 1.00$ and $n_1 = 1.55$); this implies that the γ_3 parameter has a high value, and the exponential decay of the field in that region makes it attenuate over short distances in the x-direction. Finally, we can observe that in the four corners, corresponding to the dotted areas in Figure 3.34,

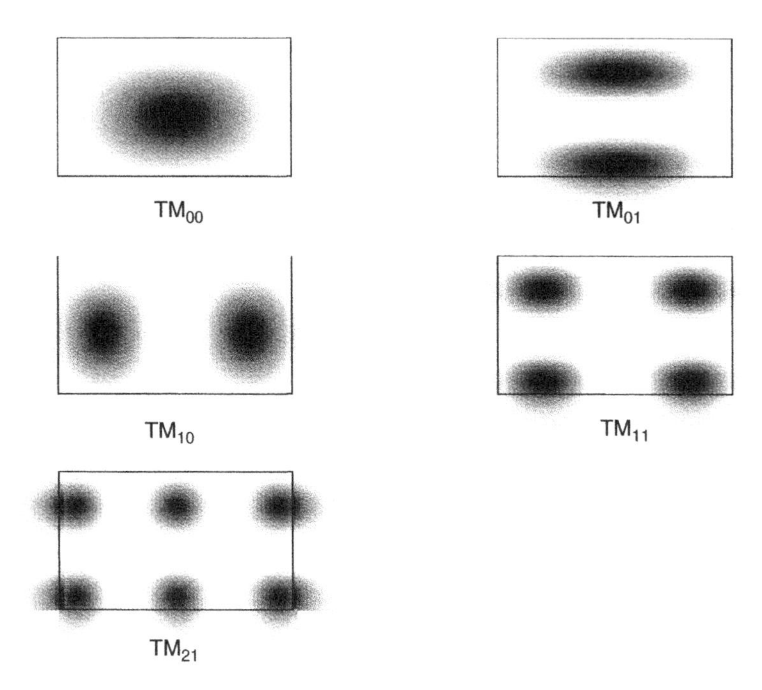

Figure 3.35 Intensity distribution for the quasi-TM modes (E^x_{pq} polarised modes) supported by a rectangular channel waveguide with dimensions $a = 2$ μm and $b = 3$ μm. The refractive index of the five homogeneous regions are: $n_1 = 1.55$, $n_2 = 1.45$, $n_3 = 1.00$, $n_4 = 1.45$ and $n_5 = 1.45$, the wavelength being 1.3 μm

the energy is very low, thus confirming the validity of the initial approximation made for the applicability of this method.

E^y_{pq} modes The solutions for quasi-TE modes (E^y_{pq} polarised modes) are obtained in a similar manner to that carried out for quasi-TM modes. In this case the dominant component for the electric field is E_y, while the component E_x can be ignored, and also $H_y = 0$. The longitudinal components of the fields for E^y_{pq} modes in the five regions are expressed as:

Region I

$$E_z = B\cos\kappa_x(x+\overline{\xi})\cos\kappa_y(y+\overline{\eta}) \tag{3.115}$$

$$H_z = -B(\varepsilon_0/\mu_0)^{1/2}n_1^2(\kappa_x/\kappa_y)(k/\beta)\sin\kappa_x(x+\overline{\xi})\sin\kappa_y(y+\overline{\eta}) \tag{3.116}$$

Region II

$$E_z = B(n_1^2/n_2^2)\cos\kappa_x(\overline{\xi}-a)\cos\kappa_y(y+\overline{\eta})\,exp[\gamma_2(x+a)] \tag{3.117}$$

$$H_z = -B(\varepsilon_0/\mu_0)^{1/2}n_1^2(\gamma_2/\kappa_y)(k/\beta)\cos\kappa_x(\overline{\xi}-a)\sin\kappa_y(y+\overline{\eta})\,exp[\gamma_2(x+a)] \tag{3.118}$$

Region III

$$E_z = B(n_1^2/n_3^2)\cos\kappa_x\bar{\xi}\cos\kappa_y(y+\bar{\eta})\,exp(-\gamma_3 x) \tag{3.119}$$

$$H_z = B(\varepsilon_0/\mu_0)^{1/2}n_1^2(\gamma_3/\kappa_y)(k/\beta)\cos\kappa_x\bar{\xi}\sin\kappa_y(y+\bar{\eta})\,exp(-\gamma_3 x) \tag{3.120}$$

Region IV

$$E_z = B\cos\kappa_y(b+\bar{\eta})\cos\kappa_x(x+\bar{\xi})\,exp[-\gamma_4(y-b)] \tag{3.121}$$

$$H_z = -B(\varepsilon_0/\mu_0)^{1/2}n_4^2(\kappa_x/\gamma_4)(k/\beta)\cos\kappa_y(b+\bar{\eta})\sin\kappa_x(x+\bar{\xi})\,exp[-\gamma_4(y-b)] \tag{3.122}$$

Region V

$$E_z = B\cos\kappa_y\bar{\eta}\cos\kappa_x(x+\bar{\xi})\,exp(\gamma_5 y) \tag{3.123}$$

$$H_z = B(\varepsilon_0/\mu_0)^{1/2}n_5^2(\kappa_x/\gamma_5)(k/\beta)\cos\kappa_y\bar{\eta}\sin\kappa_x(x+\bar{\xi})\,exp(\gamma_5 y) \tag{3.124}$$

After imposing the condition of continuity of the component H_z at the interfaces delimiting the region I with regions II and III (at $x = -a$ and $x = 0$), we obtain:

$$\tan(\kappa_x a) = \kappa_x(\gamma_2+\gamma_3)/(\kappa_x^2-\gamma_2\gamma_3) \tag{3.125}$$

$$\tan(\kappa_x\bar{\xi}) = \gamma_3/\kappa_x \tag{3.126}$$

On the other hand, the continuity of H_z in the regions IV and V (at $y = b$ and $y = 0$) leads to:

$$\tan(\kappa_y b) = n_1^2\kappa_y(n_5^2\gamma_4+n_4^2\gamma_5)/(n_4^2 n_5^2\kappa_y^2-n_1^2\gamma_4\gamma_5) \tag{3.127}$$

$$\tan(\kappa_y\bar{\eta}) = (n_5^2/n_1^2)(\kappa_y/\gamma_5) \tag{3.128}$$

From the transcendental equation (3.125) we obtain the parameter κ_x, and from the relation (3.127) the parameter κ_y is calculated. The phase parameters $\bar{\xi}$ and $\bar{\eta}$ are calculated following equations (3.126) and (3.128) respectively. In a similar way as before, the propagation constants of the quasi-TE modes are finally obtained from the relation given in (3.114). The transversal components of the electric and magnetic fields are calculated by differentiation of the equations (3.83)–(3.86). Following this procedure, the modal propagation constants and their associated electromagnetic fields remain completely characterised.

Nevertheless, as was already commented, Marcatili's method is only valid for rectangular channel waveguides having homogeneous regions, and for guided modes far from cut-off condition. If one is interested in analysing a channel waveguide with a different geometry, this method is not useful, and it is necessary to turn to others approximate methods, such as the effective index method.

3.6.2 The effective index method

The effective index method (EIM) is an approximate analysis for calculating the propagation modes of channel waveguides. It applies the tools developed for planar waveguides to solve the problem of two-dimensional structures [16]. This method is one of the simplest approximate methods for obtaining the modal fields and the propagation constant in channel waveguides having arbitrary geometry and index profiles. It consists of solving the problem in one dimension, described by the x coordinate, in such a way that the other coordinate (the y coordinate) acts as a parameter. In this way, one obtains a y-dependent effective index profile; this generated index profile is treated once again as a one-dimensional problem from which the effective index of the propagating mode is finally obtained.

The propagation constants supported by a 2D channel waveguide having a refractive index profile which depends on two coordinates $n = n(x, y)$ are then calculated by solving the propagation modes for two 1D planar waveguides. The EIM treats the channel waveguide as the superimposition of two 1D waveguides: planar waveguide I confines the light in the x direction, while planar waveguide II traps the light in the y direction (Figure 3.36). For propagating modes polarised mainly along the x direction (E^x_{pq}), we have seen that the major field components are E_x, H_y and E_z. The propagation of these polarised modes is similar to the TM modes in a 1D planar waveguide, and their solutions will correspond to the effective indices N_{I}. Now the second planar waveguide (waveguide II) is considered to be built from a guiding film of refractive index N_{I}, which has previously been calculated. The modes for the second planar waveguide are TE polarised, with E_x, H_y and H_z as non-vanishing components, because the light is mainly polarised along the x direction.

Let us consider the two-dimensional scalar wave equation for modes propagating in a channel structure (equation (3.88)), defined by its refractive index function given by $n(x, y)$:

$$\frac{\partial^2 E(x, y)}{\partial x^2} + \frac{\partial^2 E(x, y)}{\partial y^2} + [k_0^2 n^2(x, y) - \beta^2]E(x, y) = 0 \tag{3.129}$$

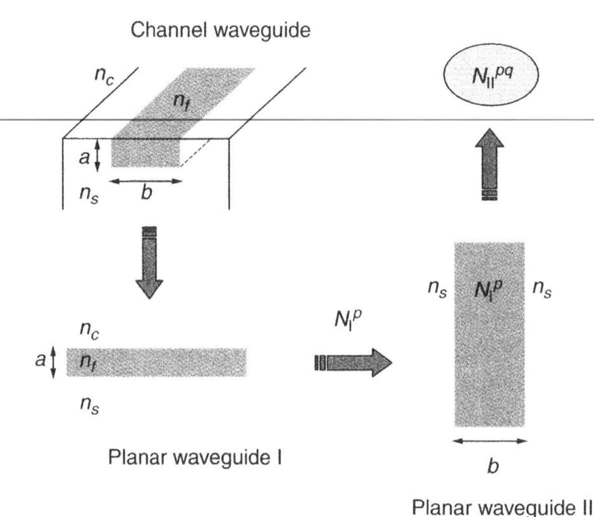

Figure 3.36 Scheme of the effective index method for solving the propagation constant of a step-index channel waveguide. Starting from a 2D waveguide, the problem is split into two step-index planar waveguides

where $k_0 = 2\pi/\lambda$, λ is the wavelength of the radiation, and β is the propagation constant of the mode, related to the effective refractive index N by $\beta = k_0 N$. The EIM is based on the assumption that the function describing the optical field admits a factorisation of the form:

$$E(x, y) = \Theta(x, y)\Phi(y) \tag{3.130}$$

Substituting this proposed solution into the wave equation for channel waveguides (3.129), we obtain a system of two coupled differential equations:

$$\frac{\partial^2 \Theta(x, y)}{\partial x^2} + [k_0^2 n^2(x, y) - k_0^2 N(y)^2]\Theta(x, y) = 0 \tag{3.131}$$

$$\frac{\partial^2 \Phi}{\partial y^2} - \left(\frac{2}{\Theta}\frac{\partial \Theta}{\partial y}\right)\frac{\partial \Phi}{\partial y} + \left(k_0^2 N^2(y) + \frac{1}{\Theta}\frac{\partial^2 \Theta}{\partial y^2} - \beta^2\right)\Phi = 0 \tag{3.132}$$

The first step in the EIM procedure consists of solving the differential equation (3.131), using the y coordinate as a parameter. This equation is similar to the wave equation for planar waveguide given in (3.26). The eigenvalue solution of equation (3.131) gives an effective index profile $N(y)$, which depends explicitly on the y coordinate. Once the index distribution $N(y)$ has been obtained, we introduce this function in the second differential equation (3.132), thus performing the second step in the problem resolution.

The approximation used by the effective index method consists of assuming that the function $\Theta(x, y)$ has a slowly variation respect to the y coordinate [17]. Thus the terms in the differential equation (3.132) that involve partial derivatives of the function $\Theta(x, y)$ respect the coordinate y (second and fourth terms) can be ignored. This approximation leads to:

$$\frac{\partial^2 \Phi}{\partial y^2} + [k^2 N^2(y) - \beta^2]\Phi = 0 \tag{3.133}$$

In this way from equation (3.132) we have obtained a decoupled differential equation, that besides that, is similar to the wave equation for planar waveguides. Then the solution of equation (3.133) finally provides the effective index of the propagating mode in the channel waveguide. This effective index is defined by two integer numbers p and q, reflecting the pth and qth order solution of equations (3.131) and (3.133) respectively.

In order to see how the EIM operates in practice, let us consider a rectangular channel waveguide similar to the one shown in Figure 3.36. The depth and width of the core waveguide are a and b respectively. The waveguide core with refractive index n_f is embedded in a substrate of refractive index of n_s being the upper part delimited by the cover with refractive index n_c.

Starting from the channel waveguide we first build an asymmetric step-index planar waveguide, which consists of a film of width a and refractive index n_f, surrounded by a cover and a substrate, having refractive indices n_c and n_s respectively. This is the planar waveguide I (Figure 3.36). The effective indices supported by this waveguide can be calculated by using any of the methods seen in Section 3.4 to solve step-index planar structures. The effective index associated to the pth guided mode for this waveguide is denoted by N_I^p.

Now the second step consists of considering a symmetric step-index planar waveguide (waveguide II) formed by a core film of thickness b, whose refractive index is the effective refractive index N_{I}^{p} calculated previously. The film is surrounded on both sides by a medium with refractive index equal to the substrate refractive index n_s. This new waveguide can easily be solved by conventional methods applied to planar structures. The effective refractive index N_{II}^{pq} of the q order guided mode calculated by this planar waveguide corresponds to the modal effective index for the channel waveguide.

The effective index method can be extended in a natural way to channel waveguides with an arbitrary refractive index profile $n = n(x, y)$, such as the structure shown in Figure 3.37. This is the typical index profile found in channel waveguides fabricated by diffusion methods: the index increase is maximum at the surface, and it decreases monotonically both in depth and laterally, until it reaches the substrate value.

Following the method outlined in Figure 3.38, for a fixed y coordinate, for instance say y_i, the index distribution as a function of the depth is given by $n = n(x, y_i)$, which defines a planar waveguide with graded-index profile. This planar waveguide can be

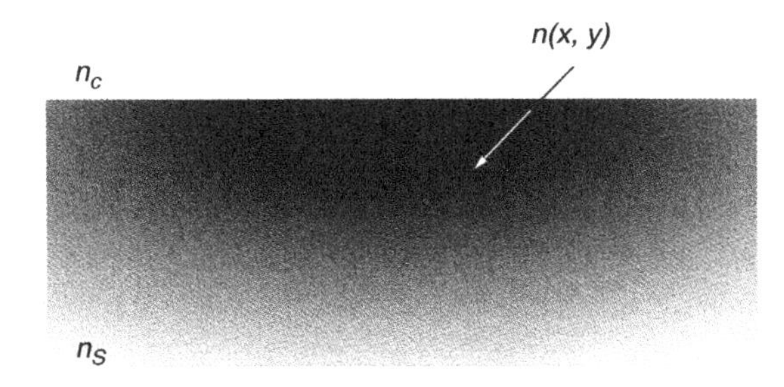

Figure 3.37 Refractive index profile of a graded-index channel waveguide. The index increase shows Gaussian-like profile in both directions, typical in diffused waveguides

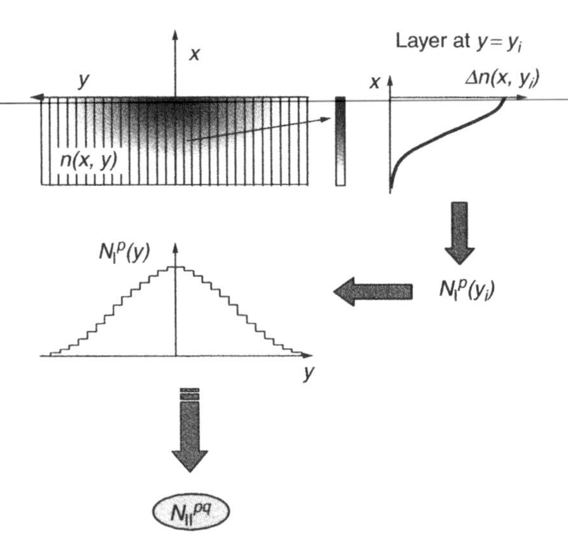

Figure 3.38 Scheme of the effective index method for calculating the propagating modes in graded-index channel waveguides

solved by the aforementioned methods, and its solution is a discrete set of effective indices. Let us assume that we choose an arbitrary, but fixed, mode order p among this set of refractive indices, and we denote it by $N_{\mathrm{I}}^{p}(y_i)$. By solving each planar waveguide for each value of the y coordinate, a one-dimensional distribution of refractive indices $N_{\mathrm{I}}^{p}(y)$ is obtained, which defines a symmetric planar waveguide, with graded-index profile along the y-direction. The propagation modes for this second 1D waveguide can be calculated, resulting in a new set of effective indices N_{II}^{pq}, where q indicates the mode order. The value of N_{II}^{pq} obtained denotes the effective index of pq mode order corresponding to the channel waveguide.

When the effective index distribution $N_{\mathrm{I}}^{p}(y)$ is calculated as a TE solution in the first planar waveguide, the propagation constant for the TM solution corresponding to the effective index profile of the qth mode is the approximate propagation constant of the quasi-TE_{pq} mode for the channel waveguide.

As an illustration of the method, let us consider the case of a graded-index channel waveguide in lithium niobate, fabricated by metallic diffusion. This fabrication process induces a refractive index increase of the substrate in the region where the diffusion takes place: the maximum index is located at the surface, and the index increase has a Gaussian distribution in depth and also laterally. The function that describes this profile is given by:

$$n(x, y) = n_s + \Delta n \cdot e^{-(x^2/a^2 + y^2/b^2)} \tag{3.134}$$

Figure 3.39 (left, above) shows a transversal view of the diffused channel waveguide with the mentioned index profile, defined by the following parameters:

$$n_s = 2.1675$$

$$\Delta n = 0.013$$

$$a = 3\ \mu\mathrm{m}$$

$$b = 5\ \mu\mathrm{m}$$

Table 3.2 summarises the results provided by the effective index method applied to the graded-index channel waveguide for quasi-TM modes calculated for a wavelength of $\lambda = 0.8\ \mu\mathrm{m}$. The values for quasi-TE modes are very similar, because the index increase with respect to the substrate index is very small ($\Delta n \ll n_s$).

The modal intensity distributions for each of the confined modes that supports the channel waveguide are represented in Figure 3.39, besides the two-dimensional index profile $n(x, y)$. It can be observed that the modes are symmetric with respect to the y direction, but not in the x-direction, thus reflecting the symmetry defined by the refractive index profile $n(x, y)$. The radiation is very well confined for the TM_{00} mode, but this confinement is reduced for higher order modes, in such a way that the

Table 3.2 Effective refractive indices for quasi-TM modes at 0.8 μm supported by the graded-index channel waveguide shown in Figure 3.39

	0	1	2	3
0	2.1736	2.1715	2.1696	2.1681
1	2.1679	—	—	—

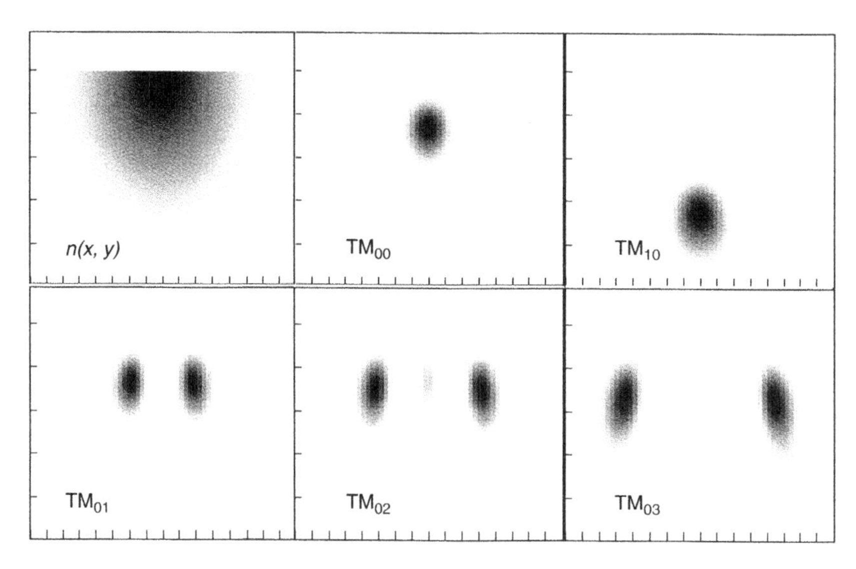

Figure 3.39 Intensity distribution for the quasi-TM modes supported by the graded-index channel waveguide defined by $n(x, y)$ (above, left), calculated at $\lambda = 0.8$ μm. Each tic corresponds to one micron. The waveguide parameters are indicated in the text

TM_{10} mode's energy is poorly concentrated. In fact, this mode has an effective index of 2.1679, which is very close to that of the substrate value (2.1675), indicating that the mode is very close to the cut-off. Finally, let us note that the p and q modal indices correspond to the number of zeros in the intensity distribution along the x and y direction respectively.

Notes

1. The electromagnetic waves defined by equations (3.12) and (3.13) are clearly non-planar waves, because of the spatial dependence of the fields' amplitudes. Indeed, a planar wave must have an infinite spatial extension, and this is not the case when dealing with optical waveguides modes that have transversal dimensions of the order of the wavelength of the light, typically of a few microns.
2. The longitudinal components of the fields are not independent, but related by:

$$\beta \frac{\partial H_z}{\partial x} = \varepsilon_0 n^2 \omega \frac{\partial E_z}{\partial y}$$

References

[1] G. Lifante, F. Cussó, F. Meseguer and F. Jaque, "Solar Concentrators using Total Internal Reflection", *Applied Optics*, **22**, 3966–3970 (1983).

[2] R. Stoffer, H.J.W.M. Hoekstra, R.M. de Ridder, E. Van Groesen and F.P.H. Van Beckum, "Numerical Studies of 2D Photonic Crystals: Waveguides, Coupling Between Waveguides and Filters", *Optical and Quantum Electronics*, **32**, 947–961 (2000).

[3] G. Lifante, T. Balaji and A. Muñoz-Yagüe, "Planar Optical Waveguides Fabricated by Molecular Beam Epitaxy of Pd-Doped CaF_2 Layers", *Applied Physics Letters*, **70**, 2079–2081 (1997).

[4] R. Nevado and G. Lifante, "Low-Loss, Damage-Resistant Optical Waveguides in Zn-Diffused $LiNbO_3$ by a Two-Step Procedure", *Applied Physics A*, **72**, 725–728 (2001).

[5] P.L. Pernas, E. Ruiz, J.L. Cantaño, J. Garrido and B.J. García, "Channel Waveguides Grown by Selective Area Beam Epitaxy", *Optical Materials*, **17**, 259–262 (2001).

[6] P.L. Pernas, M.J. Hernández, E. Ruíz, E. Cantelar, R. Nevado, C. Morant, G. Lifante and F. Cussó, "Zn-Vapor Diffused Er:Yb:$LiNbO_3$ Channel Waveguides Fabricated by Means of SiO_2 Electron Cyclotron Resonance Plasma Deposition", *Applied Surface Science*, **161**, 123–130 (2000).

[7] A. Ghatak and K. Thyagarajan, *Introduction to Fiber Optics*, Cambridge University Press, Cambridge (1998).

[8] N. Nishihara, M. Haruna and T. Suhara, *Optical Integrated Circuits*, Mac-Graw Hill, R.R. Donelley & Sons Company, United States of America (1989).

[9] R.H. Dicke and J.P. Wittke, *Introduction to Quantum Mechanics*, Reading, MA: Addison-Wesley (1960), pp. 245–248.

[10] J.M. White and F. Heidrich, "Optical Waveguide Refractive Index Profiles Determined from Measurement of Mode Indices: A Simple Analysis", *Applied Optics*, **15**, 151–155 (1976).

[11] K.S. Chiang, "Construction of Refractive-Index Profiles of Planar Dielectric Waveguides from the Distribution of Effective Indexes", *Journal of Lightwave Technology*, **LT-3**, 385–391 (1985).

[12] W.H. Press, S.A. Tevkolsky, W.T. Vetterling and B.P. Flannery, *Numerical Recipes in Fortran 77*, Chapter 3. Cambridge University Press, New York (1996).

[13] M. Kawachi, "Silica Waveguides on Silicon and Their Application to Integrated-Optic Components", *Optical and Quantum Electronics*, **22**, 391–416 (1990).

[14] R.C. Alferness, "Titanium-Diffused Lithium Niobate Waveguide Devices", in Guided-Wave Optoelectronics, T. Tamir (ed.), Springer-Verlag, Berlin (1990).

[15] E.A.J. Marcatili, "Dielectric Rectangular Waveguide and Directional Coupler for Integrated Optics", *The Bell System Technical Journal*, **48**, 2071–2102 (1969).

[16] K.S. Chiang, "Dual Effective-index Method for the Analysis of Rectangular Waveguides", *Applied Optics*, **25**, 2169–2174 (1986).

[17] J. Liñares, V. Moreno and M.C. Nistal, "Designing of Monomode Step-Index Channel Guides with Quasi-Exact Modal Solutions by the Effective Index Method", *Journal of Modern Optics*, **47**, 599–604 (2000).

4

COUPLED MODE THEORY: WAVEGUIDE GRATINGS

Introduction

We have shown in the last chapter that a z-invariant optical structure that defines an optical waveguide admits solutions in the form of modes derived from Maxwell's equations, whose z-dependence is expressed by the simple function $exp(-i\beta z)$. In this chapter we will see that there is an orthogonal relation between the modes supported by a waveguide structure, which implies that each mode propagates independently along the waveguide, and therefore there is no power exchange between modes. Moreover, the modes form a complete set of orthogonal functions in the sense that any arbitrary electromagnetic field in a waveguide can be expressed as a superimposition of the waveguide modes.

Nevertheless, energy transfer between modes can take place if the original waveguide structure is altered; this means that the modes (originally independent) become mutually "coupled". The theory that describes the interaction between modes, or the new modes that arise from the modified structure, is known as the *coupled mode theory* (CMT). The main aim of this theory is to derive the coupling coefficients that drive the optical power exchange between the modes of the original structure.

We will apply the results of the CMT to one of the most important integrated photonics elements: the waveguide grating [1]. Waveguide diffraction gratings are periodic perturbations induced in planar or channel waveguides, and are widely used in integrated photonics technology to perform a great variety of functions: input/output couplers, reflectors, modal conversors, etc. Also, gratings are optical elements of high wavelength dispersion, enabling their application in wavelength filters, separators/combiners and multiplexors/demultiplexors [2]. We will describe the two main types of waveguide gratings, and obtain the coupling coefficients in both cases, discussing the fundamental phenomena involved in mode coupling induced by waveguide diffraction gratings.

4.1 Modal Coupling

4.1.1 Modal orthogonality and normalisation

Let us consider a dielectric loss-less waveguide (planar or channel) having an arbitrary geometry, but which is invariant in the propagation direction z. This means that the structure can be described by the spatial distribution of the dielectric permittivity of the form $\varepsilon = \varepsilon(x, y)$ (Figure 4.1). The electromagnetic field corresponding to monochromatic radiation propagating along this structure must fulfil Maxwell's equations:

$$\nabla \times \mathbf{E} = -i\omega\mu_0\mathbf{H} \tag{4.1}$$

$$\nabla \times \mathbf{H} = i\omega\varepsilon\mathbf{E} \tag{4.2}$$

Let us now consider two electromagnetic waves described by their complex field amplitudes $(\mathbf{E}_1, \mathbf{H}_1)$ and $(\mathbf{E}_2, \mathbf{H}_2)$. As these complex amplitudes must fulfil the above equations, the following relation can be deduced [3] (see Appendix 4):

$$\nabla(\mathbf{E}_1 \times \mathbf{H}_2^* + \mathbf{E}_2^* \times \mathbf{H}_1) = 0 \tag{4.3}$$

From the relation (4.3) the following expression can be obtained (see Appendix 5):

$$\int_S \frac{\partial}{\partial z}[\mathbf{E}_1 \times \mathbf{H}_2^* + \mathbf{E}_2^* \times \mathbf{H}_1]_z \, dx\, dy = 0 \tag{4.4}$$

As the dielectric permittivity $\varepsilon = \varepsilon(x, y)$ which defines the waveguiding structure does not depend on the z coordinate, Maxwell's equations admit mode solutions, whose z dependence can be expressed by the simple function $exp(-i\beta z)$. These solutions are referred to as normal modes of the z-invariant structure, where β is the modal propagation constant. If the two electromagnetic waves in equation (4.4) correspond to two arbitrary normal modes ν and μ, the complex amplitudes for the electric and magnetic fields $(\mathbf{E}_1, \mathbf{H}_1)$ and $(\mathbf{E}_2, \mathbf{H}_2)$ should be expressed as:

$$\mathbf{E}_1(x, y, z) = \mathbf{E}_\nu(x, y)e^{-i\beta_\nu z} \tag{4.5}$$

$$\mathbf{H}_1(x, y, z) = \mathbf{H}_\nu(x, y)e^{-i\beta_\nu z} \tag{4.6}$$

$$\mathbf{E}_2(x, y, z) = \mathbf{E}_\mu(x, y)e^{-i\beta_\mu z} \tag{4.7}$$

$$\mathbf{H}_2(x, y, z) = \mathbf{H}_\mu(x, y)e^{-i\beta_\mu z} \tag{4.8}$$

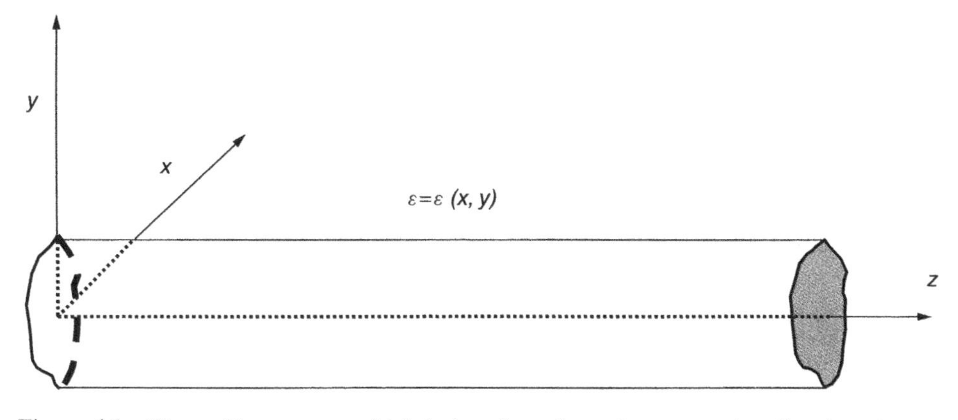

Figure 4.1 Waveguide structure which is invariant along the propagation direction, defined through the spatial distribution of its dielectric permittivity $\varepsilon(x, y)$

By substituting these expressions into equation (4.4) it yields:

$$-\iint i(\beta_v - \beta_\mu) e^{-i(\beta_v - \beta_\mu)z} \left[\mathbf{E}_v \times \mathbf{H}_\mu^* + \mathbf{E}_\mu^* \times \mathbf{H}_v \right]_z dx\, dy = 0 \tag{4.9}$$

and taking into account the vectorial relation $(\mathbf{A} \times \mathbf{B})_z = (\mathbf{A}_t \times \mathbf{B}_t)$ it follows:

$$(\beta_v - \beta_\mu) \iint \left[\mathbf{E}_{vt} \times \mathbf{H}_{\mu t}^* + \mathbf{E}_{\mu t}^* \times \mathbf{H}_{vt} \right]_z dx\, dy = 0 \tag{4.10}$$

where the subscripts t and z denote the components on the xy plane and the z component, respectively, and the integral is extended to an infinite surface S perpendicular to the z-axis. The relation in (4.10) indicates that for $\beta_\mu \neq \beta_v$ the integral must vanish: this implies that the transversal electromagnetic fields are *orthogonal* between them, and therefore the expression in (4.10) is known as the *modal orthogonality relation*.

On the other hand, and taking into account that the effective power density (in W/m^2) is given by the real part of the Poynting vector $\mathbf{S} = (1/2)\mathbf{E} \times \mathbf{H}^*$, the power flux (in watts) transported by the radiation corresponding to the modal field μ can be written as:

$$P = P_z = Re\left\{ \iint \frac{1}{2} (\mathbf{E}_t \times \mathbf{H}_t^*)_z \, dx\, dy \right\} = \frac{1}{4} \iint \left[\mathbf{E}_{\mu t} \times \mathbf{H}_{\mu t}^* + \mathbf{E}_{\mu t}^* \times \mathbf{H}_{\mu t} \right]_z dx\, dy \tag{4.11}$$

From the above expression it is worth noting that the power flux P takes positive and negative values for $\beta_\mu > 0$ and $\beta_\mu < 0$ respectively; this is so because the positive and negative values of the propagation constant β_μ correspond to waves propagating in the positive and negative direction of the z axis, respectively. If the power flux P is set to 1 W in equation (4.11), the electric and magnetic fields associated with a mode are said to be *normalised*.

By combining the orthogonality relation (4.10) with the *power flux normalisation* (equation (4.11) with $P = 1$ W), we obtain the expression known as the *modal orthonormalisation*, given by:

$$\frac{1}{4} \iint \left[\mathbf{E}_{vt} \times \mathbf{H}_{\mu t}^* + \mathbf{E}_{\mu t}^* \times \mathbf{H}_{vt} \right]_z dx\, dy = \pm \delta_{\mu v} \tag{4.12}$$

where we have assumed that the relation is established for guides modes, having discrete values of β, and $\delta_{\mu v}$ denotes the *Kronecker's delta* function. For radiation modes, which have a continuum spectrum on β, this relation takes the form:

$$\frac{1}{4} \iint \left[\mathbf{E}_{vt} \times \mathbf{H}_{\mu t}^* + \mathbf{E}_{\mu t}^* \times \mathbf{H}_{vt} \right]_z dx\, dy = \pm \delta(\beta_\mu - \beta_v) \tag{4.13}$$

where $\delta(\beta_\mu - \beta_v)$ stands for the *Dirac delta* function. Obviously, the sign of the right part of equations (4.12) and (4.13) has only physical meaning for $\mu = v$, that is, when the right-hand side does not vanish. The positive sign must be taken for $\beta_\mu = \beta_v > 0$, and a minus sign for $\beta_\mu = \beta_v < 0$.

4.1.2 Modal expansion of the electromagnetic field

The electromagnetic fields of the modes in a waveguide structure form a *complete system* of orthogonal functions. This implies that any arbitrary electromagnetic field propagating along the waveguide structure can be expressed as a superposition of the modal fields. The relations of completitude and normalisation are applied to the transversal fields in the xy plane, and thus an arbitrary field can be expressed in the form:

$$\mathbf{E}_t(x, y, z) = \sum_{\nu} a_\nu \mathbf{E}_{\nu t}(x, y) e^{-i\beta_\nu z} \tag{4.14}$$

$$\mathbf{H}_t(x, y, z) = \sum_{\nu} a_\nu \mathbf{H}_{\nu t}(x, y) e^{-i\beta_\nu z} \tag{4.15}$$

where the *expansion coefficient* a_ν represents the weighted contribution of the νth mode to the electromagnetic field, and the symbol Σ denotes summation over all the modes. Let us note that the expansion coefficients are a-dimensional magnitudes.

As the modes in a waveguide structure include guided modes with a discrete spectrum, as well as radiation modes with a continuum spectrum, the expansions in equations (4.14) and (4.15) imply a discrete summation for guided modes and an integral on β_ν for radiation modes. The summation must be performed also for different polarisations and for each degenerate mode. It is also important to recall that, for a given mode, β can take positive as well as negative values (corresponding to waves travelling in $+z$ and $-z$ directions). Therefore, these two equations should include positive and negative terms on β_ν.

When the integral Σ denotes the above described superposition, the transversal fields $\mathbf{E}_t$ and $\mathbf{H}_t$ can be fully expressed using the relations (4.14) and (4.15). In addition, the z-component of the fields $\mathbf{E}_z$ and $\mathbf{H}_z$ can be calculated by Maxwell's equations:

$$\nabla_t \times \mathbf{E}_t = -i\omega\mu_0 \mathbf{H}_z \tag{4.16}$$

$$\nabla_t \times \mathbf{H}_t = i\omega\varepsilon \mathbf{E}_z \tag{4.17}$$

Thus, any arbitrary field $(\mathbf{E}, \mathbf{H})$ is completely determined by a set of modal expansion coefficients a_ν.

The waveguide structure that defines the modal fields $\{\mathbf{E}_\nu, \mathbf{H}_\nu\}$ is known as the *canonical structure*. As each *canonical mode* (*normal mode*) is a solution of Maxwell's equations for this structure, each mode can exist independently. In addition, the a_ν coefficients which define an arbitrary electromagnetic field propagating along the canonical structure are constant. This implies that there is *no interaction* between modes, or in other words, there cannot exist energy transfer or *coupling* between modes.

In a waveguide structure slightly different to the canonical structure, the normal modes cannot propagate independently, because these modes are *not* solutions of Maxwell's equation for the new structure. Nevertheless, the modal field can still be expressed as a superposition of normal modes using relations (4.14) and (4.15), although in this case the a_ν coefficients are no longer constant but are a function of the propagation distance z:

$$\mathbf{E}_t(x, y, z) = \sum_{\nu} a_\nu(z) \mathbf{E}_{\nu t}(x, y) e^{-i\beta_\nu z} \tag{4.18}$$

$$\mathbf{H}_t(x, y, z) = \sum_{\nu} a_\nu(z) \mathbf{H}_{\nu t}(x, y) e^{-i\beta_\nu z} \tag{4.19}$$

The consequence of this is that the modes of the new structure can only be expressed as a modal superposition, the modal amplitude coefficients a_ν being dependent on the propagation distance as a result of *mode coupling*. The spatial evolution of the a_ν coefficients due to mode coupling is driven by the so-called *mode-coupling equations*, and the theory describing this behaviour is known as the coupled mode theory (CMT) [4].

4.1.3 Coupled mode equations: coupling coefficients

Let us assume that $\mathbf{E}^{(0)}$ and $\mathbf{H}^{(0)}$ represent the electric and magnetic field amplitudes associated with an arbitrary wave propagating along a canonical waveguide structure, described by a given distribution of the dielectric permittivity $\varepsilon = \varepsilon(x, y)$. As these fields must fulfil Maxwell's equations, equations (4.1) and (4.2) become:

$$\nabla \times \mathbf{E}^{(0)} = -i\omega\mu_0\mathbf{H}^{(0)} \tag{4.20}$$

$$\nabla \times \mathbf{H}^{(0)} = i\omega\varepsilon\mathbf{E}^{(0)} \tag{4.21}$$

We now consider a new structure, which is a modification of the canonical structure. It can be described by $\varepsilon + \Delta\varepsilon = \varepsilon(x, y) + \Delta\varepsilon(x, y)$, $\Delta\varepsilon$ being the change in the dielectric permittivity of the modified structure respect to the canonical one. Any arbitrary electromagnetic field ($\mathbf{E}, \mathbf{H}$) in the new structure must satisfy:

$$\nabla \times \mathbf{E} = -i\omega\mu_0\mathbf{H} \tag{4.22}$$

$$\nabla \times \mathbf{H} = i\omega(\varepsilon + \Delta\varepsilon)\mathbf{E} \tag{4.23}$$

The spatial distribution $\Delta\varepsilon(x, y)$ which describes the difference between the modified and the canonical structures can be considered as a modification or a perturbation of the canonical structure. Nevertheless, at this point the magnitude of $\Delta\varepsilon$ has not to be necessarily small compared to ε.

Following a similar procedure to that given in Appendix 4, by combining equations (4.20) and (4.21) with equations (4.22) and (4.23), we obtain an equation that relates the fields of the canonical structure to the fields of the modified structure, coupled through the change in the permittivity:

$$\nabla(\mathbf{E} \times \mathbf{H}^{(0)*} + \mathbf{E}^{(0)*} \times \mathbf{H}) = -i\omega\mathbf{E}^{(0)*}\Delta\varepsilon\mathbf{E} \tag{4.24}$$

By making an integration of this expression over a cylindrical volume with an infinitely large area parallel to the xy plane and with an infinitesimal width in the z direction, as is shown in Figure 4.2, it results in:

$$\iint \frac{\partial}{\partial z}\left[\mathbf{E}_t \times \mathbf{H}_t^{(0)*} + \mathbf{E}_t^{(0)*} \times \mathbf{H}_t\right]_z dx\,dy = -i\omega \iint \mathbf{E}^{(0)*}\Delta\varepsilon\mathbf{E}\,dx\,dy \tag{4.25}$$

where we have proceeded in a similar way to that shown in Appendix 5, except that the right part of this equation is not zero because of the contribution of $\Delta\varepsilon$.

If we assume further that the fields $\mathbf{E}^{(0)}$ and $\mathbf{H}^{(0)}$ correspond to the μth mode of the canonical structure, their transversal components can be expressed as:

$$\mathbf{E}_t^{(0)}(x, y, z) = \mathbf{E}_{\mu t}(x, y)e^{-i\beta_\mu z} \tag{4.26}$$

$$\mathbf{H}_t^{(0)}(x, y, z) = \mathbf{H}_{\mu t}(x, y)e^{-i\beta_\mu z} \tag{4.27}$$

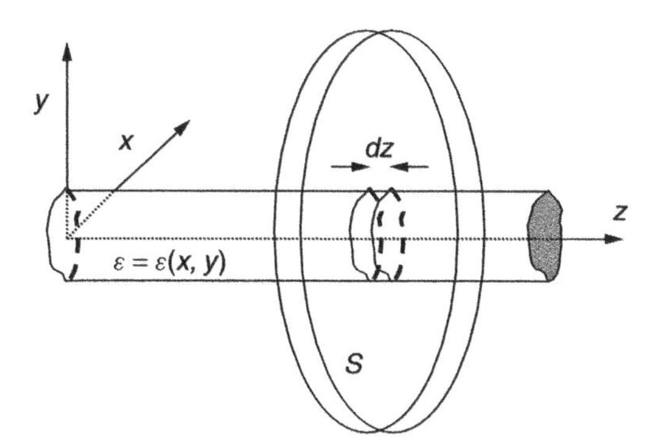

Figure 4.2 Volume taken for performing the integration of equation (4.24)

On the other hand, the modal expansion given by equations (4.14) and (4.15) can be used to obtain the transversal field amplitudes $\mathbf{E}_t$ and $\mathbf{H}_t$. In this way, the left-hand side (*LHS*) of equation (4.25) leads to:

$$LHS = \sum_{\nu} \frac{\partial}{\partial z} \left\{ a_\nu(z) e^{-i(\beta_\nu - \beta_\mu)z} \iint \left[\mathbf{E}_{\nu t} \times \mathbf{H}^*_{\mu t} + \mathbf{E}^*_{\mu t} \times \mathbf{H}_{\nu t} \right]_z dx\, dy \right\} \tag{4.28}$$

The integral in the above expression, taking into account the orthonormalisation relation (4.12), yields $\pm 4\delta_{\mu\nu}$, and thus the *LHS* reduces to:

$$LHS = \pm 4 \frac{\partial a_\mu(z)}{\partial z} = \pm 4 \frac{d a_\mu(z)}{dz} \tag{4.29}$$

where the + and − signs must be chosen for positive and negative values of β_μ, respectively.

The complete expression for the electric field complex amplitude can be given as a function of the field modal expansion given in (4.18) and (4.19), yielding:

$$\begin{aligned} \mathbf{E} &= \mathbf{E}_t + \mathbf{E}_z = \mathbf{E}_t + \nabla_t \times \mathbf{H}_t / i\omega(\varepsilon + \Delta\varepsilon) \\ &= \sum_\nu a_\nu \left[\mathbf{E}_{\nu t} + \nabla_t \times \mathbf{H}_{\nu t} / i\omega(\varepsilon + \Delta\varepsilon) \right] e^{-i\beta_\nu z} \\ &= \sum_\nu a_\nu \left[\mathbf{E}_{\nu t} + \frac{\varepsilon}{\varepsilon + \Delta\varepsilon} \mathbf{E}_{\nu z} \right] e^{-i\beta_\nu z} \end{aligned} \tag{4.30}$$

where, in addition, we have used a relation similar to that (4.17) to obtain $\mathbf{E}_z$. Taking into account this result, the right-hand side (*RHS*) of equation (4.25) becomes:

$$RHS = -i\omega \sum_\nu a_\nu e^{-i(\beta_\nu - \beta_\mu)z} \iint \mathbf{E}^*_\mu \Delta\varepsilon \left[\mathbf{E}_{\nu t} + \frac{\varepsilon}{\varepsilon + \Delta\varepsilon} \mathbf{E}_{\nu z} \right] dx\, dy \tag{4.31}$$

By combining the results obtained in (4.29) and (4.31), equation (4.25) can finally be written in the form:

$$\pm \frac{d a_\mu(z)}{dz} = -i \sum_\nu \kappa_{\mu\nu}(z) a_\nu(z) e^{-i(\beta_\nu - \beta_\mu)z} \tag{4.32}$$

where we have introduced the *coupling coefficient* $\kappa_{\mu\nu}(z)$ defined as:

$$\kappa_{\mu\nu}(z) \equiv \kappa^t_{\mu\nu}(z) + \kappa^z_{\mu\nu}(z) \tag{4.33}$$

$$\kappa^t_{\mu\nu}(z) \equiv \frac{\omega}{4} \iint \mathbf{E}^*_{\mu t}(x, y)\Delta\varepsilon(x, y, z)\mathbf{E}_{\nu t}(x, y)\, dx\, dy \tag{4.34}$$

$$\kappa^z_{\mu\nu}(z) \equiv \frac{\omega}{4} \iint \mathbf{E}^*_{\mu z}(x, y)\frac{\varepsilon\Delta\varepsilon(x, y, z)}{\varepsilon + \Delta\varepsilon(x, y, z)}\mathbf{E}_{\nu z}(x, y)\, dx\, dy \tag{4.35}$$

which has the dimension of m^{-1}.

Equation (4.32), which includes summation over all the ν modes (guided as well as radiation modes), represents the *coupling modal equation*, and determines the spatial variation (along the propagation distance z) of the modal amplitude coefficients a_μ. The parameter $\kappa_{\mu\nu}(z)$ is the coupling coefficient between modes μ and ν, and has been expressed as a summation of $\kappa^t(z)$ representing the coupling contribution due to the transversal component of the electric field, and $\kappa^z(z)$ that involves the coupling between the longitudinal component of the fields. These coupling coefficients are evaluated by substituting the normalised electric field profiles into equations (4.34) and (4.35), and making the integration in those spatial regions where $\Delta\varepsilon$ is different to zero.

In the expressions of the coupling coefficients we have explicitly included their z-dependence, because we allow the altered structure to be z-dependent through the modification of the dielectric permittivity $\Delta\varepsilon(x, y, z)$. This is indeed the most common case, for instance, in mode coupling induced by periodic structures [1].

Let us note that if $\Delta\varepsilon$ is a real quantity (non-absorbing structures), from (4.34) and (4.35), it follows that the coupling coefficients must fulfil the relation:

$$\kappa_{\nu\mu}(z) = \kappa^*_{\mu\nu}(z) \tag{4.36}$$

On the other hand, the power flux in the propagation direction z along the waveguide structure is given by the summation of the modal power fluxes $\pm a^2$, and the derivative on z can be expressed by equation (4.32) as:

$$\frac{d}{dz}\sum_\mu |a_\mu(z)|^2 = \sum_\mu \left(a_\mu \frac{da^*_\mu}{dz} + a^*_\mu \frac{da_\mu}{dz}\right) = -i\sum_{\mu\nu}(\kappa_{\mu\nu} - \kappa^*_{\nu\mu})a^*_\mu a_\nu e^{-i(\beta_\nu-\beta_\mu)z} \tag{4.37}$$

Taking into account the relation (4.36), the right-hand side of this equation is zero. This result implies that the power flux is maintained as the wave propagates along the waveguide structure. Therefore, the *power flux conservation* can be expressed in terms of the coupling coefficients by:

$$\frac{d}{dz}\sum_\mu |a_\mu(z)|^2 = 0 \tag{4.38}$$

Let us note that if the structure presents absorption (or gain, in the case of lasers and amplifiers), $\Delta\varepsilon$ is complex, and the relation (4.38) is no longer valid.

It is important to note here that in order to obtain the coupling modal equations it has not been necessary to make any approximation. This means that if the superposition of all the modes is done for Σ_ν, the electromagnetic wave evolution is adequately described by equation (4.32). Nevertheless, an analysis using all the modes is not practical in most cases, because (4.32) represents a set of equations of infinite dimension, which should include radiation modes. Therefore, when the modal coupling theory is applied in practice, only a small number of modes are considered, which are chosen taking into account reasonable physical conditions, such as in the case of phase matching conditions. Using this procedure, equation (4.32) is reduced to a small number of equations. In most cases only two guided modes are considered, and this is the analysis that we will be following in next section. In this approximate analysis it is necessary to choose an appropriate canonical structure in such a way that the structure to be analysed can be represented by a $\Delta\varepsilon$ as small as possible, because in this case the accuracy of the approximation will depend on the magnitude of $\Delta\varepsilon$. In order to achieve that, one can choose a canonical structure in which $\Delta\varepsilon = 0$ except for a small region in the structure.

Finally, let us consider the approximate expressions for the coupling coefficients [5]. The longitudinal components z of the fields corresponding to guided modes are usually much smaller that the transversal components, and therefore it holds that:

$$|\kappa^t| \gg |\kappa^z| \tag{4.39}$$

Using this result, in a first approximation the κ^t coefficient can be used as the coupling coefficient κ, thus ignoring the contribution of the coefficient κ^z. Even if κ^z is not ignored, making use of the fact that $\varepsilon \gg \Delta\varepsilon$, equation (4.35) can be approximated to:

$$\kappa^z_{\mu\nu}(z) \approx \frac{\omega}{4} \iint \mathbf{E}^*_{\mu z}(x, y)\Delta\varepsilon(x, y, z)\mathbf{E}_{\nu z}(x, y)\, dx\, dy \tag{4.40}$$

and therefore, from this equation and equation (4.34) one finally obtains:

$$\kappa_{\mu\nu}(z) \approx \frac{\omega}{4} \iint \mathbf{E}^*_{\mu}(x, y)\Delta\varepsilon(x, y, z)\mathbf{E}_{\nu}(x, y)\, dx\, dy \tag{4.41}$$

as an approximate expression for the coupling coefficient $\kappa_{\mu\nu}(z)$.

So far, we have discussed the coupling coefficients in isotropic media, where the coupling between mutually orthogonal polarised modes is not possible. This implies, for example, that coupling between TE–TM or TM–TE modes in planar waveguides cannot take place in isotropic structures. In the general case of *anisotropic media*, the polarisation induced in the medium by the electric field of the light does not have the same direction as the electric vector, and this implies that the constitutive relation (2.13) involving the displacement vector and the electric field $\mathcal{D} = \varepsilon\mathcal{E}$ has to be substituted by a tensorial relation. This leads to the possibility of coupling between modes that have orthogonal polarisations. As a matter of fact, in anisotropic media the dielectric permittivity is not longer a scalar, but a tensor, and the constitutive relation becomes:

$$\mathcal{D} = \varepsilon\mathcal{E} \tag{4.42}$$

where the dielectric permittivity ε is given by:

$$\varepsilon = \begin{pmatrix} \varepsilon_{xx} & \varepsilon_{xy} & \varepsilon_{xz} \\ \varepsilon_{yx} & \varepsilon_{yy} & \varepsilon_{yz} \\ \varepsilon_{zx} & \varepsilon_{zy} & \varepsilon_{zz} \end{pmatrix} \tag{4.43}$$

If the change on the dielectric permittivity induced by the perturbation in the original canonical structure $\Delta\varepsilon$ have non-null off-diagonal elements, the coupling between perpendicular polarised modes can take place. In the general case, the change in the dielectric permittivity tensor can be expressed as:

$$\Delta\varepsilon = \begin{pmatrix} \Delta\varepsilon_{xx} & \Delta\varepsilon_{xy} & \Delta\varepsilon_{xz} \\ \Delta\varepsilon_{yx} & \Delta\varepsilon_{yy} & \Delta\varepsilon_{yz} \\ \Delta\varepsilon_{zx} & \Delta\varepsilon_{zy} & \Delta\varepsilon_{zz} \end{pmatrix} \tag{4.44}$$

Although the coupling coefficient in the case of anisotropic induced changes in the dielectric permittivity is rather complicated, it can be demonstrated that if this change is small ($\Delta\varepsilon \ll \varepsilon$), a compact formula can be derived [3], resulting in:

$$\kappa_{\mu\nu}(z) \approx \frac{\omega}{4} \iint \mathbf{E}_{\mu}^{*}(x, y)\Delta\varepsilon(x, y, z)\mathbf{E}_{\nu}(x, y)\, dx\, dy \tag{4.45}$$

where the whole vectors of the electric mode fields enter in this expression, and not just the transverse or longitudinal part as it was in the case of isotropic media. In many practical applications, the difference between the perturbed dielectric tensor and the dielectric tensor corresponding to the canonical structure is small, and then the expression given in (4.45) provides a good approximation for the coupling coefficient in practical problems.

4.1.4 Coupling mode theory

As we have seen in previous sections, a loss-less waveguide structure which is invariant along the propagation direction (z axis) can support several guided modes, which are defined by the waveguide structure and its boundary conditions. There is an orthogonality relation between modes, in such a way that they can propagate independently, without mutual coupling and without energy transfer between them.

In contrast, if the waveguide is slightly altered, for instance, by inducing a small change in the refractive index in a region close to the waveguide, the original modes of the unperturbed waveguide are no longer independent, but they will be *mutually coupled*. There are two methods for analysing the optical propagation of the modes in the perturbed waveguide: (i) one can calculate the normal modes of the new structure by solving Maxwell's equations with appropriate boundary conditions; and (ii) it is possible to express the perturbed optical field as a summation of the normal modes corresponding to the unperturbed waveguide [6]. Of course, the first method gives exact solutions, but the resolution of the problem is sometimes very difficult. On the other hand, although the second procedure provides only approximate solutions, it is simple and direct. Also, this method allows a qualitative comprehension of the essential features and phenomena involved in the problem, and the solution given is usually quite accurate. Therefore, the *mode coupling theory* is a method that can be used for the optical propagation description in a perturbed waveguide by means of the known normal modes corresponding to the unperturbed structure.

Although we have seen that the optical field is described by a discrete summation of all the guided modes plus the continuum radiation modes, in many cases it suffices to consider only two guided modes; in other words, only two coefficients a_ν in the modal expansion are different to zero.

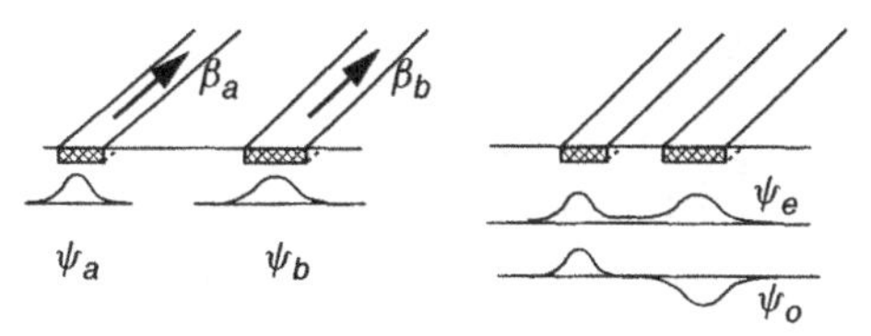

Figure 4.3 Left: structure of two independent waveguides (uncoupled), having modal fields Ψ_a and Ψ_b, and propagation constants β_a and β_b. Right: when the two waveguides approach the modal fields modify to Ψ_e and Ψ_o, as a result of the mutual coupling

The behaviour of two modes having mutual coupling can be examined by considering the situation depicted in Figure 4.3. The unperturbed waveguide (canonical structure) consists of two waveguides, I and II, which are invariant along the propagation direction z, and each waveguide support a single guided mode, a and b. The separation distance between the waveguides is large enough so that there is no mutual influence, in such a way that each mode can propagate independently. The modal field distributions corresponding to the guided modes a and b are given by Ψ_a and Ψ_b, respectively, and their propagation constants are β_a and β_b, respectively. In our discussion we assume that $\beta_a < \beta_b$.

Let us now examine the modification suffered by the normal modes when the two waveguides are uniformly coupled along the propagation direction, which can be done by reducing the separation distance between them.

When the two waveguides approach each other, the initial canonical structure is perturbed: the original normal modes (canonical modes) Ψ_a and Ψ_b are no longer modes of the new structure, and instead two new normal modes Ψ_e and Ψ_o appear, which propagate along the coupled waveguide system. The propagation constants of these two modes are β_e and β_o, and it holds that $\beta_e > \beta_b$ and $\beta_o < \beta_a$. As the fields Ψ_e and Ψ_o are now the normal modes of the new structure, they can be excited separately, and can also propagate independently.

Also, under appropriate light injection, the two modes Ψ_e and Ψ_o can be excited simultaneously. Since the propagation constants of the modes are different, they propagate at different speeds, giving rise to a beat phenomenon: if one looks at the behaviour of the light on the waveguides I and II, it seems that the optical power transfers periodically between the two waveguides, in such a way that the energy is apparently transferred from mode Ψ_a to mode Ψ_b and vice versa, as the optical perturbation proceeds along the waveguide system. This coupling effect is stronger as the values of propagation constants β_a and β_b become closer, as is shown in Figure 4.4 (different waveguides) and Figure 4.5 (identical waveguides), where the optical power in the waveguide system is plotted as a function of the propagation distance, after light injection into waveguide I.

This behaviour describes the main features and fundamental concepts of modal coupling behaviour between propagating modes in waveguides structures. The purpose of the modal coupling theory is to obtain the modal fields Ψ_e and Ψ_o, the propagation constants β_e and β_o, the beat period, and other relevant parameters, as a function of the known modes Ψ_a and Ψ_b corresponding to the unperturbed system.

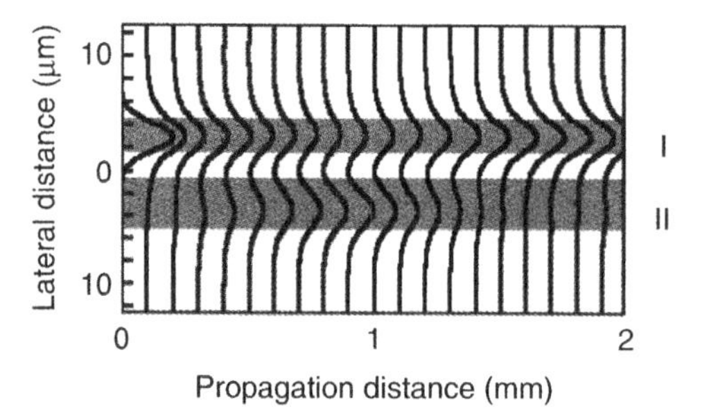

Figure 4.4 System of two coupled waveguides. Both waveguides are monomode, but have different propagation constants. After launching light into waveguide I, there is a partial power exchange into waveguide II

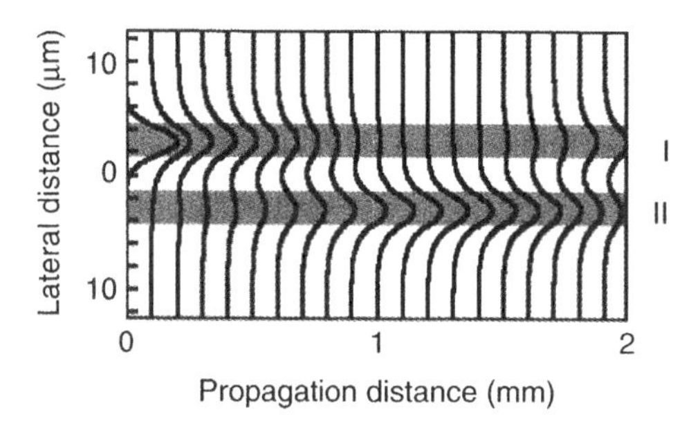

Figure 4.5 System of two coupled monomode waveguides, having the same value of their propagation constants. After launching light into waveguide I, there is a total power exchange into waveguide II

The electromagnetic field of a monochromatic wave propagating along the waveguide structure can usually be well represented by the main component of its associated electric field (or magnetic field). We will refer to this component as the *optical field*, and thus our description is based on a scalar description of the optical propagation. If Ψ_a and Ψ_b describe the optical fields associated with the guided modes of the coupled waveguide system I and II, they can be expressed as:

$$\Psi_a(x, y, z, t) = A(z)e^{-i\beta_a z} f_a(x, y)e^{i\omega t} \tag{4.46}$$

$$\Psi_b(x, y, z, t) = B(z)e^{-i\beta_b z} f_b(x, y)e^{i\omega t} \tag{4.47}$$

where $f_a(x, y)$ and $f_b(x, y)$ are the field distribution functions which have been normalised to the power flux over the transversal section of the waveguide system. If the coupling between waveguide I and waveguide II is reduced to zero by separating sufficiently the waveguides, the fields Ψ_a and Ψ_b are reduced to the original independent normal modes, and the amplitudes $A(z)$ and $B(z)$ are reduced to constant values.

On the other hand, if the waveguides are close each other, there will exist mutual coupling, and the amplitude $A(z)$ and $B(z)$ are no longer constant, but will depend on the propagation distance z. The modal coupling equations (4.32) involving only two guided modes are reduced to:

$$\pm\frac{dA(z)}{dz} = -i\kappa_{ab}B(z)e^{-i(\beta_b-\beta_a)z} \quad (4.48)$$

$$\pm\frac{dB(z)}{dz} = -i\kappa_{ba}A(z)e^{+i(\beta_b-\beta_a)z} \quad (4.49)$$

where the coefficients κ_{ab} and κ_{ba} are the coupling coefficients between the modes a and b and vice versa, respectively. The term $exp[\pm i(\beta_b - \beta_a)z]$ corresponds to the *phase mismatching* between the two guided modes. If in equation (4.48) we set the coupling coefficients to zero ($\kappa_{ab} = \kappa_{ba} = 0$), it is easily obtained that $A(z) = A_0$ and $B(z) = B_0$ (constant values), and the optical fields Ψ_a and Ψ_b are reduced to the original optical fields.

The coupling coefficient κ_{ab} (equations (4.34) and (4.35)) takes into account the spatial overlapping of the normal mode Ψ_a over the region where the dielectric permittivity ε changes compared to the original unperturbed structure. In this case, the change on ε is restricted to waveguide II, as is shown in Figure 4.6. Therefore, the coupling coefficient κ_{ab} is calculated according to the following integral:

$$\kappa_{ab} = C\int_{II} f_a^*\Delta\varepsilon f_b\,dx\,dy \quad (4.50)$$

where the integration range extends over the transversal section of waveguide II, which is where the perturbation $\Delta\varepsilon$ felt by waveguide I occurs. The constant C which appears in (4.50) is determined by taking into account the normalisations for the functions Ψ_a and Ψ_b.

The behaviour of the optical fields modified by the coupling effect can be determined by calculating the propagation constants of the waveguide system by using the coupled mode equations (4.48) and (4.49). If the modes are collinear, two cases can be distinguished in the coupled field behaviour: when the two waves have the same propagation direction (unidirectional coupling), and when the coupled waves have opposite directions (bi-directional coupling). Indeed, we will show that for the existence of this second type of coupling, it is necessary to introduce an additional structure to the waveguide system (for instance, a periodic modulation of the refractive index) in such a way that the modes can have an efficient coupling.

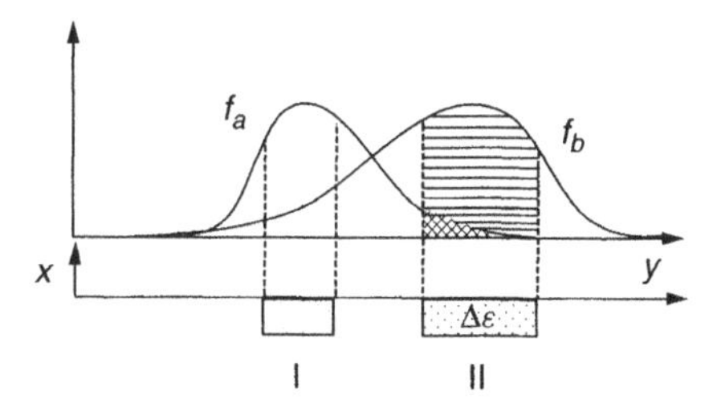

Figure 4.6 Coupled waveguides, showing the modal field distributions of the uncoupled waveguides. Dashed area denotes the region where the integration in equation (4.50) takes place, which is used for obtaining the coupling coefficient κ_{ab}

4.1.5 Co-directional coupling

In this case we assume that two collinear guided modes propagate along the coupled waveguides system in the same direction ($+z$ direction), and thus the propagation constants associated to the guided modes have positive values: $\beta_a > 0$, $\beta_b > 0$. The coupled modal equations in this case are written as:

$$\frac{dA(z)}{dz} = -i\kappa B(z)e^{-i(\beta_b-\beta_a)z} \tag{4.51}$$

$$\frac{dB(z)}{dz} = -i\kappa A(z)e^{+i(\beta_b-\beta_a)z} \tag{4.52}$$

where it holds that $\kappa_{ab} = \kappa_{ba}^* = \kappa$, and κ is a real and positive magnitude. The coupled modal equations (4.51)–(4.52) in fact constitute a set of two *differential coupled equations*, linked through the coupling coefficient κ. To solve these differential equations we postulate solutions of the form:

$$A(z) = Ae^{-i\gamma z}e^{-i\Delta z} \tag{4.53}$$

$$B(z) = Be^{-i\gamma z}e^{+i\Delta z} \tag{4.54}$$

where we have introduced a new magnitude Δ, defined by:

$$2\Delta \equiv \beta_b - \beta_a \tag{4.55}$$

This parameter, called the *mismatching* and expressed in m^{-1}, gives idea of the *degree of synchronism* between the modes a and b, and will allow us to discuss the favourable conditions for the existence of efficient coupling. When the two modes are synchronised ($\beta_b = \beta_a$, or $\Delta = 0$), the situation is said to be in *perfect phase-matching* condition.

By introducing the solutions postulated in (4.53) and (4.54) into equations (4.51–4.52), we test whether they are really valid solutions providing that the parameter γ takes the value given by:

$$\gamma = \pm(\kappa^2 + \Delta^2)^{1/2} \tag{4.56}$$

The parameter γ here defined has dimensions of m^{-1}.

In addition, the relation between the A and B coefficients is given by:

$$B/A = \kappa/(\gamma - \Delta) \tag{4.57}$$

The general solutions, considering the two possible signs for the parameter γ, are finally of the form:

$$A(z) = \left(A_e e^{-i\gamma z} + A_o e^{+i\gamma z}\right)e^{-i\Delta z} \tag{4.58}$$

$$B(z) = \left(\frac{\kappa A_e}{\gamma - \Delta}e^{-i\gamma z} - \frac{\kappa A_o}{\gamma + \Delta}e^{+i\gamma z}\right)e^{+i\Delta z} \tag{4.59}$$

The complete expressions for the modal fields in waveguides I and II are obtained by inserting the $A(z)$ and $B(z)$ coefficients into equations (4.46)–(4.47), resulting in:

$$\Psi_a(x, y, z, t) = [A_e e^{-i\beta_e z} + A_o e^{-i\beta_o z}]f_a(x, y)e^{i\omega t} \tag{4.60}$$

$$\Psi_b(x, y, z, t) = [B_e e^{-i\beta_e z} + B_o e^{-i\beta_o z}]f_b(x, y)e^{i\omega t} \tag{4.61}$$

where the quotient between A_e and B_e has a positive constant value and the ratio between A_o and B_o is a negative constant.

By examining the expressions (4.58)–(4.59), the new propagation constants β_e and β_o of the coupled waveguides are then determined by:

$$\beta_e = (\beta_a + \beta_b)/2 + \gamma \tag{4.62}$$

$$\beta_o = (\beta_a + \beta_b)/2 - \gamma \tag{4.63}$$

where we have chosen the + sign in the formula (4.56) for the parameter γ.

From these formulae we can deduce that the value for the propagation constant β_e is the semi-sum of β_a and β_b plus the constant γ, and the propagation constant β_o is this semi-sum minus γ. Moreover, the new propagation constants fulfil the relations $\beta_e > \beta_a, \beta_b$ and $\beta_o < \beta_a, \beta_b$. Figure 4.7 shows these results graphically, indicating how the propagation constants change as result of the waveguides coupling.

Figure 4.8 shows the propagation constant values corresponding to a waveguide system which consists of two identical monomode step-index planar waveguides, as a function of their separation distance t. When the waveguides are separated by a large enough distance ($t > 15$ μm), their propagation constants are equal (synchronous waveguides). As the separation distance decreases, the propagation constants are modified according to Figure 4.7. The plot in Figure 4.8 includes the result obtained by the coupled mode theory, as well as the propagation constant obtained by exact calculation by solving Maxwell's equations for the coupled system. We observe that the CMT gives good results for weak coupling, up to a separation distance of $t \sim 5$ μm. As the distance between the waveguides is reduced, the values provided by the CMT differ from the exact values since the coupling is increased.

An example of the modification suffered by the propagation constants and field profiles of two monomode waveguides as they approach each other is presented in Figure 4.9 (different monomode waveguides) and Figure 4.10 (identical waveguides). In Figure 4.9 we observe that when the waveguides are well separated, the propagation constants and the field profiles correspond to those of two isolated monomode waveguides, and modes 0 and 1 correspond to the wide and narrow waveguide respectively. As the waveguides approach, the propagation constants modify: the lower propagation constant (mode 1) decreases and the higher propagation constant increases its value (mode 0). Also, the field profiles change appreciably when the waveguides interact,

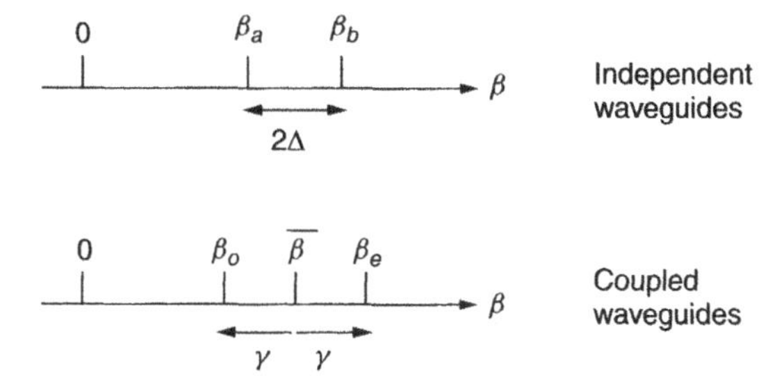

Figure 4.7 Scheme representing the propagation constant values of the guided modes for two independent waveguides (above) and the new propagation constants of the modes when the waveguides are coupled (below), as a result of a spatial approach between them

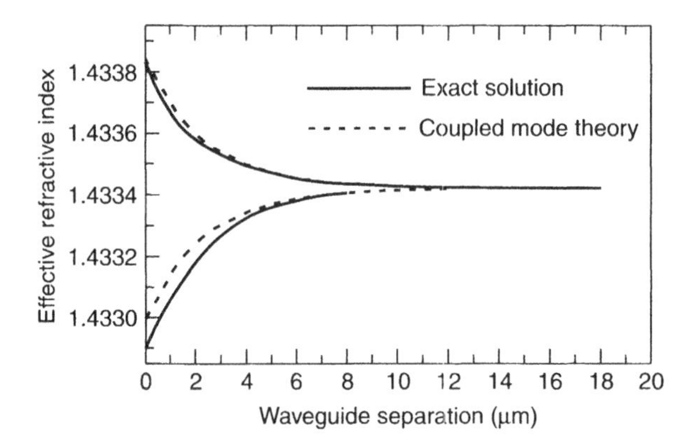

Figure 4.8 Comparison between the mode effective indices (propagation constants, $\beta = k_0 n_{\text{eff}}$) provided by the CMT (dashed line) and the exact values (continuous line) as function of the waveguides, separation. Waveguide parameters: identical step-index symmetric waveguides; $d = 3$ μm; $n_c = 1.43280$, $n_f = 1.43423$; $\lambda = 0.633$ μm

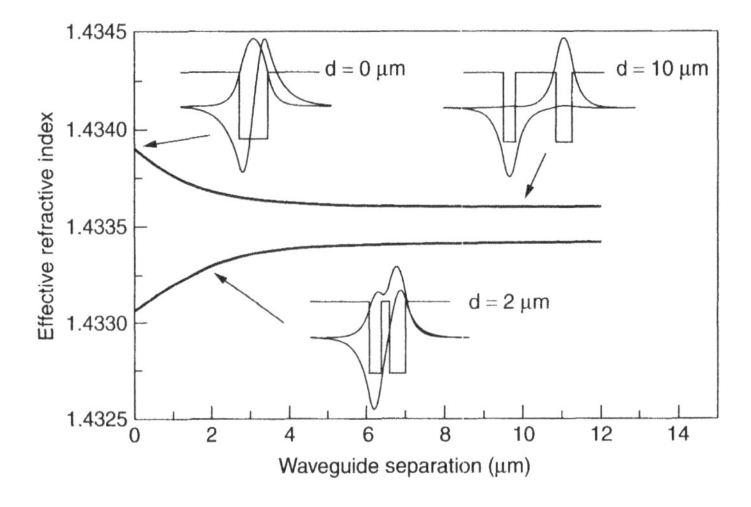

Figure 4.9 Evolution of the propagation constants and modal field profiles of two monomode waveguides as the separation distance reduces. Waveguide parameters: step-index symmetric waveguides; $d_1 = 3$ μm; $d_2 = 4$ μm; $n_c = 1.43280, n_f = 1.43423$; $\lambda = 0.633$ μm

and they can be considered to be built by a linear combination of the two modal field profiles corresponding to the isolated waveguides.

In the case of identical waveguides (Figure 4.10) the propagation constants are equal for non-interacting waveguides ($t > 3$ μm), but as the waveguides reduce their separation distance the propagation constant splits in two different values, following the relation (4.62)–(4.63). Also, the field profiles are constructed by a linear combination of the modal field profiles for independent waveguides: the mode 0 becomes $\Psi_e = \Psi_a + \Psi_b$, and the mode 1 $\Psi_o = \Psi_a - \Psi_b$.

Equations (4.60)–(4.61) indicate that the waves Ψ_a and Ψ_b propagating along the mutually coupled waveguides I and II can be expressed as a linear combination of

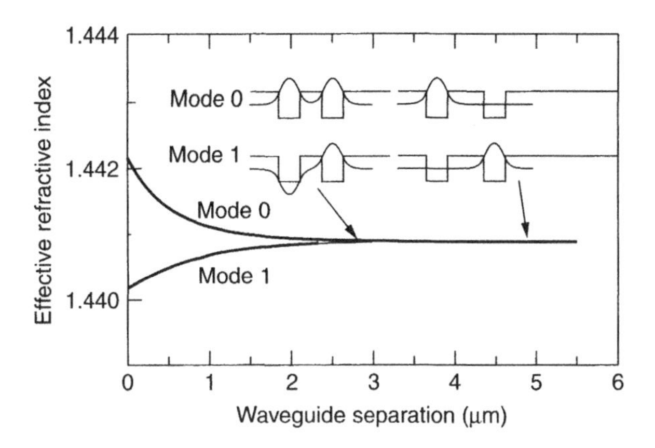

Figure 4.10 Propagation constants and modal field profiles of two identical waveguides as a function of their separation distance. Waveguide parameters: step-index symmetric waveguides; $d = 3$ μm; $n_c = 1.43280$, $n_f = 1.44283$; $\lambda = 0.633$ μm

two waves with propagation constants β_e and β_o. These two propagation constants are therefore the propagation constants of the coupled waveguide system above described, whose wave descriptions are given by:

$$\Psi_e = A_e f_a + B_e f_b \tag{4.64}$$

$$\Psi_o = A_o f_a + B_o f_b \tag{4.65}$$

where A_e and A_o are constants determined by the initial condition of light injection.

Let us now examine the particular case in which waveguide I is selectively excited at the input at $z = 0$. In this case, we have that $A(0) = 1$ and $B(0) = 0$ at $z = 0$, and the coefficients $A(z)$ and $B(z)$ are described by:

$$A(z) = e^{-i\Delta z}\left[\cos\gamma z + i\frac{\Delta}{\gamma}\sin\gamma z\right] \tag{4.66}$$

$$B(z) = -e^{+i\Delta z}\frac{i\kappa}{\gamma}\sin\gamma z \tag{4.67}$$

As f_a and f_b are normalised functions, the power fluxes along waveguides I and II are described by $|A(z)|^2$ and $|B(z)|^2$, respectively. In this way, equations (4.66) and (4.67) can be written in terms of the flux power as:

$$\frac{|A(z)|^2}{|A(0)|^2} = 1 - F\sin^2\gamma z \tag{4.68}$$

$$\frac{|B(z)|^2}{|A(0)|^2} = F\sin^2\gamma z \tag{4.69}$$

where for convenience we have defined the a-dimensional magnitude F as:

$$F \equiv \left(\frac{\kappa}{\gamma}\right)^2 = \frac{1}{1 + (\Delta/\kappa)^2} \tag{4.70}$$

Figure 4.11 shows the curves calculated using the above equations, where it can be seen that the power of the two waves propagating along the z direction has a periodical variation. Following equations (4.68) and (4.69), F $(0 \leqslant F \leqslant 1)$ indicates the maximum transfer of optical power that can take place between waveguides I and II. The maximum power transfer is achieved for a distance $z = L$ for which it holds that $\sin^2 \gamma z = 1$, that is, $\gamma z = m\pi/2$, and then the minimum distance that fulfils this condition is given by:

$$L = \pi/2\gamma \tag{4.71}$$

This distance is called the *coupling length.*

If the two modes are synchronised ($\beta_a = \beta_b$, $\Delta = 0$, $\gamma = \kappa$), the F value is maximum ($F = 1$) and the coupling length in this case is reduced to:

$$L = \pi/2\kappa \tag{4.72}$$

Figure 4.12 shows a system of two parallel and identical monomode waveguides (at $\lambda = 0.633$ μm) separated by a distance of 1 μm, with the parameters given in the figure caption. This coupled system corresponds to the case where the two interacting modes are synchronised. As we can seen, after launching light into the left waveguide, a total power transfer occurs after a propagation distance of ~500 μm. This distance can be directly calculated by computing the coupling coefficient using the CMT, and the results are given in Figure 4.13. We can observe that for a separation distance between the waveguides of 1 μm the CMT gives a coupling length of ~500 μm, in accordance with the result shown in Figure 4.12.

From the analysis described above, two main features can be highlighted: first, when the two modes are synchronised ($\beta_a = \beta_b$) the coupling coefficient has only influence on the coupling length (see equation (4.72)), and a power transfer of 100% is always obtained, regardless of the coupling coefficient κ; second, when the modes are not synchronised ($\beta_a \neq \beta_b$), a 100% transfer is not possible, and the degree of transfer is determined by the coupling coefficient κ and the degree of synchronisation Δ.

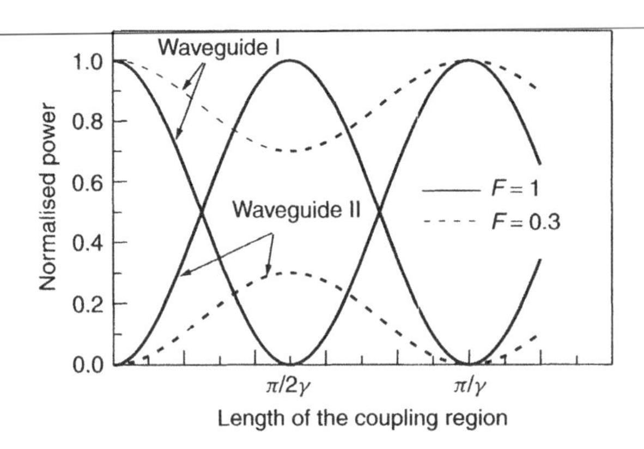

Figure 4.11 Variation of the relative optical power in waveguides I and II as a function of the propagation distance, in the case of synchronous waveguides ($F = 1$, continuous line) and asynchronous waveguides with $\Delta \neq 0$ ($F = 0.3$, dashed line). In both cases the maximum power transfer is achieved at $L = \pi/2\gamma$

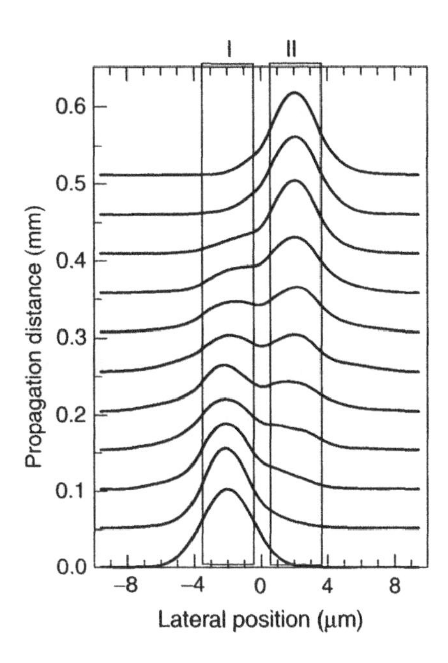

Figure 4.12 Transversal intensity profiles corresponding to two parallel planar waveguides, separated by a distance of 1 μm, after the injection of light at $z = 0$ into waveguide I. The step-index symmetric waveguides are identical, and have the following parameters: $n_c = 1.4328$; $n_f = 1.4342$; film thickness $d = 3$ μm; wavelength $\lambda = 0.633$ μm. Each uncoupled waveguide has a single TE guided mode with $\beta = 14.23$ μm^{-1}. For a propagation distance of $z = 0.5$ mm, all the power is transferred to waveguide II. This distance is the coupling length, and corresponds to that calculated by using the CMT, as indicated in Figure 4.13

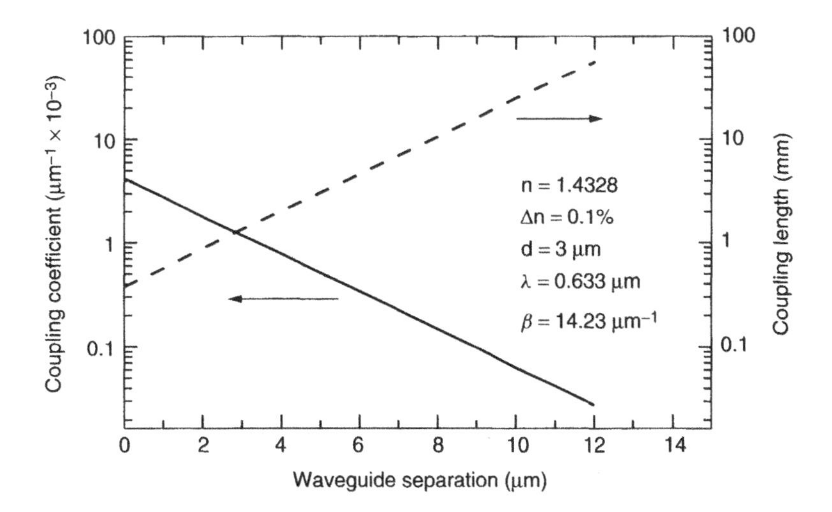

Figure 4.13 Computation of the coupling coefficient κ (continuous line) corresponding to the TE guided modes for two parallel monomode waveguides, as a function of their separation distance, calculated by the CMT. Dashed line represents the coupling length L. The waveguide system has the same parameters as those indicated in Figure 4.12

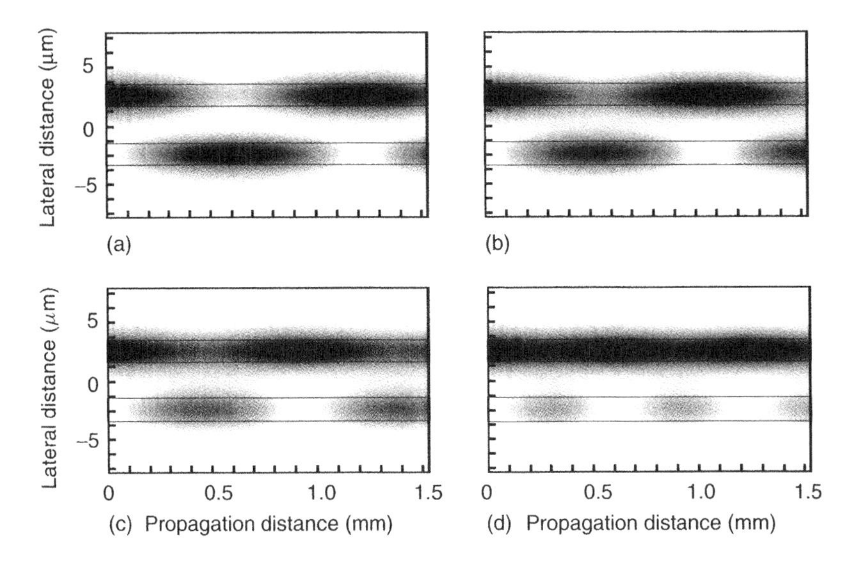

Figure 4.14 Light propagation in a waveguide system which consists of two parallel step-index waveguides, separated by a fixed distance. The upper waveguide has a fixed width of $d_1 = 1.6$ μm. The lower waveguide has a width of: (a) $d_1 = 1.6$ μm; (b) $d_2 = 1.7$ μm; (c) $d_2 = 1.8$ μm and (d) $d_2 = 1.9$ μm

In order to see the influence of the mismatching parameter on the power transfer and on the coupling length, we have plotted in Figure 4.14 the optical field intensity (indicated by a grey scale) in a system of two parallel monomode waveguides. In all four cases, waveguide I, where the light is launched at $z = 0$, has the same width ($d_1 = 1.6$ μm). Case (a) corresponds to identical waveguides, and so the second waveguide has a width of $d_2 = 1.6$ μm. This is the case of synchronised modes, where it can be observed that a total transfer occurs. If the two waveguides are not identical (from (b) $d_2 = 1.7$ μm, (c) $d_2 = 1.8$ μm to (d) $d_2 = 1.9$ μm), two features can be outlined. First, as the mismatching between the two guides increases, the fraction of power transfer decreases. In fact, for a width $d_2 = 1.9$ μm, the power transferred into waveguide II is very low, and almost all the energy remains in waveguide I. A second effect observed as the mismatching increases is the shortening in the coupling length. According to equation (4.56), for a given coupling coefficient κ, the parameter γ increases with the mismatching 2Δ, and following the formula given in (4.71) the coupling length decreases, in agreement with the behaviour found on the sequence of pictures in Figure 4.14.

4.1.6 Contra-directional coupling

Let us now consider the case in which collinear coupling takes place between a mode propagating in the $+z$ direction and a mode propagating in the $-z$ direction, that is, two contrapropagating modes, where their propagation constants will have opposite signs. We assume that the mode a propagates in the positive direction, and thus $\beta_a > 0$, and mode b propagates in the $-z$ direction, $\beta_b = -|\beta_b| < 0$. Under these circumstances the coupling between contrapropagating modes is not possible, regardless of the waveguides' separation. For coupling to take place it is necessary to have an additional element that induces coupling between the contrapropagating modes, and this can be done for instance by introducing a periodic perturbation in the refractive index in the region between the waveguides [7], as shown in Figure 4.15.

Let us assume that the coupling coefficient induced by the periodic perturbation can be expressed as:

$$\kappa_{ab} = \kappa e^{-i\delta z} \tag{4.73}$$

where κ is a real positive quantity. From equations (4.48) and (4.49), we obtain the coupling modal equations involving only two guided modes, and coupled via the coupling coefficient given in (4.73):

$$\frac{dA(z)}{dz} = -i\kappa B(z) e^{-i(-|\beta_b|-\beta_a+\delta)z} \tag{4.74}$$

$$\frac{dB(z)}{dz} = +i\kappa A(z) e^{+i(-|\beta_b|-\beta_a+\delta)z} \tag{4.75}$$

From these equations it is clear that the modes a and b cannot be synchronised because of the different signs of the propagation constants β_a and β_b. Nevertheless, this synchronisation can be attained with $-|\beta_b| + \delta$ thanks to the presence of the periodical perturbation, which adds an extra contribution δ to the propagation constants. In this case, the parameter that characterised the *degree of synchronism* is given by:

$$2\Delta = -|\beta_b| - \beta_a + \delta \tag{4.76}$$

Figure 4.16 shows schematically the relation between the propagation constants β_a and β_b, the parameter δ corresponding to the periodic structure, and the *mismatch parameter* 2Δ.

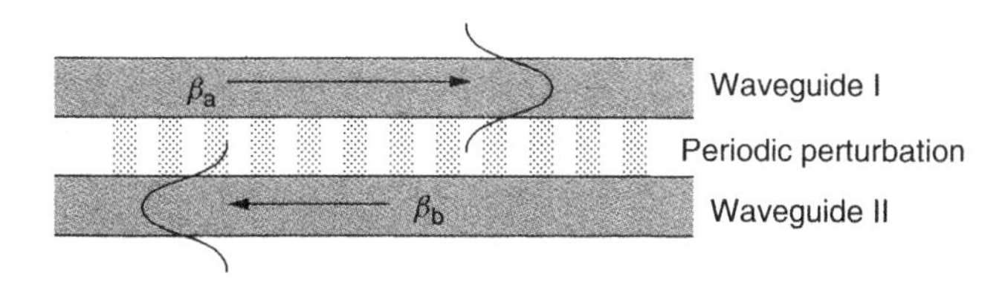

Figure 4.15 Periodic perturbation in the region between waveguides I and II, that allows optical power transfer between contrapropagating modes

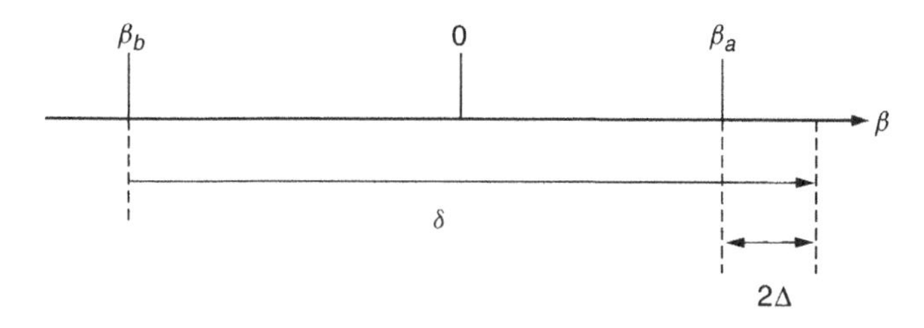

Figure 4.16 Scheme that shows the synchronisation of contrapropagating modes in the presence of a periodic structure

For solving the coupled differential equations (4.74) and (4.75) we postulate solutions of the form:

$$A(z) = Ae^{-i\gamma z}e^{-i\Delta z} \tag{4.77}$$

$$B(z) = Be^{-i\gamma z}e^{+i\Delta z} \tag{4.78}$$

By introducing these expressions into equations (4.74) and (4.75), it follows that (4.77) and (4.78) are valid solutions providing that the parameter γ is given by:

$$\gamma = \pm i\sqrt{\kappa^2 - \Delta^2} \tag{4.79}$$

where the coefficients A and B must fulfil the relation:

$$B/A = \kappa/(\Delta - \gamma) \tag{4.80}$$

Finally, the complete expressions for the coefficients $A(z)$ and $B(z)$ remain as:

$$A(z) = \left(A_+e^{+\alpha z} + A_-e^{-\alpha z}\right)e^{-i\Delta z} \tag{4.81}$$

$$B(z) = \left(\frac{\kappa A_+}{\Delta - i\alpha}e^{+\alpha z} + \frac{\kappa A_-}{\Delta + i\alpha}e^{-\alpha z}\right)e^{+i\Delta z} \tag{4.82}$$

where the parameter α has been defined by:

$$\alpha \equiv -i\gamma = \sqrt{\kappa^2 - \Delta^2} \tag{4.83}$$

and A_+ and A_- are coefficients which should be determined by the initial conditions of light injection for each particular case.

The complete solution for the optical fields ψ_a and ψ_b is obtained by inserting the solution for $A(z)$ and $B(z)$ into equations (4.46)–(4.47), yielding finally:

$$\Psi_a(x, y, z, t) = [A_+e^{+\alpha z} + A_-e^{-\alpha z}]e^{-i(\beta_a+\Delta)z}f_a(x, y)e^{i\omega t} \tag{4.84}$$

$$\Psi_b(x, y, z, t) = \left[\frac{\kappa A_+}{\Delta - i\alpha}e^{+\alpha z} + \frac{\kappa A_-}{\Delta + i\alpha}e^{-\alpha z}\right]e^{-i(\beta_b-\Delta)z}f_b(x, y)e^{i\omega t} \tag{4.85}$$

In conditions of synchronism or quasi-synchronism, where $|\Delta| \ll \kappa$, the value of the parameter α defined in (4.83) is a real quantity. The first and second terms in equation (4.84) correspond to the two normal modes of the coupled waveguide system. One of these modes raises its amplitude with the position $+z$ along the periodic structure, while the amplitude corresponding to the other mode decreases. The modes ψ_a and ψ_b are therefore a linear combination of two waves having propagation constants $\beta_a + \Delta$ and $\beta_b - \Delta$. More, the expressions (4.84) and (4.85) indicate that the propagation constants β_1 and β_2 of the normal modes of the perturbed waveguide system (two waveguides plus the periodic structure) are complex quantities:

$$\beta_1 = \beta_a + \Delta \pm i\alpha \tag{4.86}$$

$$\beta_2 = \beta_b - \Delta \pm i\alpha \tag{4.87}$$

This implies that the modes will experience exponential attenuation or exponential amplification as they propagate along the waveguide system, depending on the contribution of the A_+ and A_- amplitude coefficients.

Let us now consider that the periodic modulation region has a length L, that is, the coupling region is restricted from $z = 0$ to $z = L$. Let us examine the behaviour of the optical power in waveguides I and II in this case, assuming that at the input ($z = 0$) we launch light into waveguide I. In this situation the initial condition gives $A(0) = 1$ at $z = 0$ and $B(L) = 0$ at $z = L$. The solutions provided by equations (4.81) and (4.82), taking into account these initial conditions, and assuming that $\kappa > \Delta$, become:

$$A(z) = e^{-i\Delta z} \frac{\alpha \cosh[\alpha(z - L)] + i\Delta \sinh[\alpha(z - L)]}{\alpha \cosh(\alpha L) - i\Delta \sinh(\alpha L)} \tag{4.88}$$

$$B(z) = e^{+i\Delta z} \frac{i\kappa \sinh[\alpha(z - L)]}{\alpha \cosh(\alpha L) - i\Delta \sinh(\alpha L)} \tag{4.89}$$

In order to find a more compact expression for the above relations, we define the a-dimensional magnitude F as:

$$F \equiv \left(\frac{\kappa}{\alpha}\right)^2 = \frac{1}{1 - (\Delta/\kappa)^2} \tag{4.90}$$

Using this new magnitude, the normalised power in waveguides I and II are given by:

$$\frac{|A(z)|^2}{|A(0)|^2} = \frac{1 + F \sinh^2[\alpha(z - L)]}{1 + F \sinh^2(\alpha L)} \tag{4.91}$$

$$\frac{|B(z)|^2}{|A(0)|^2} = \frac{F \sinh^2[\alpha(z - L)]}{1 + F \sinh^2(\alpha L)} \tag{4.92}$$

Figure 4.17 shows the curves for the relative power between waveguides I and II as function of the position along the coupling region, where the periodic perturbation exists. For a value of $\alpha = 0.2\pi/L$ it can be observed that the power transferred from waveguide I to waveguide II is ~32%; if the parameter α is set to $\alpha = 0.5\pi/L$, the power transfer is increased to a value of ~85%. It is important to note also that for positions beyond the existence of the periodic modulation ($z > L$) the power in the waveguides is not modified, because the coupling coefficient is zero and the transfer between waveguides is inhibited. Of course, the same situation is faced for position values of $z < 0$.

By examining equations (4.91) and (4.92), it can be deduced that the maximum power transfer between waveguides is achieved when $F = 1$. This condition implies that $\Delta = 0$, that is, in the *phase matching condition* given by the relation $\beta_a = -|\beta_b| + \delta$. In this case ($F = 1$, $\kappa = \alpha$), the efficiency of power transfer simplifies to:

$$\eta = \frac{|B(0)|^2}{|A(0)|^2} = \tanh^2(\kappa L) \tag{4.93}$$

Following this formula, even in the most favourable case of $F = 1$, the efficiency can never reach the unity. Nevertheless, for $\kappa L \approx \pi$ the power transfer is close to 99%, and thus an interaction region length of $L \approx \pi/\kappa$ is enough to assure a nearly total power transfer for practical purposes.

In the case of strong mismatching, when the mismatch parameter is greater than the coupling coefficient ($\Delta > \kappa$), the parameter α defined in (4.83) becomes an imaginary number, and therefore the formulae above obtained must be consequently modified. In

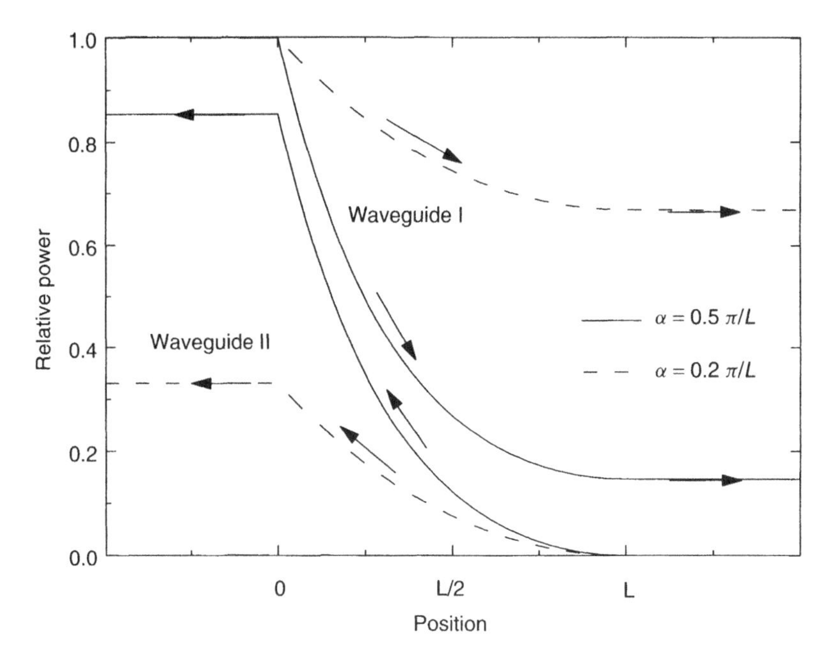

Figure 4.17 Power transfer between contrapropagating modes as a function of the position along the periodic structure of length L, for two values of the parameter α. Note that the power in the waveguides is only modified in the coupling region $0 < z < L$

particular, the hyperbolic functions become normal sinusoidal functions, and thus the formula (4.92) transforms to:

$$\frac{|B(z)|^2}{|A(0)|^2} = \frac{F \sin^2 \gamma(z - L)}{1 + F \sin^2(\gamma L)} \tag{4.94}$$

where now the parameters γ and F are given by:

$$\gamma = \sqrt{\Delta^2 - \kappa^2} \tag{4.95}$$

$$F = \left(\frac{\kappa^2}{\gamma^2}\right) \tag{4.96}$$

The behaviour of the power transfer efficiency, using (4.94) for $|\Delta| > \kappa$ and (4.92) for the range $|\Delta| < \kappa$ (setting $z = 0$), as a function of the mismatch parameter 2Δ is shown in Figure 4.18. This graph corresponds to an interaction length of $L = 300$ μm, using a coupling coefficient of $\kappa = 0.012$ μm^{-1}. We observe that for the situation close to the phase matching condition ($|\Delta| \ll \kappa$) the efficiency is very close to one, and corresponds to the value given by the formula (4.93): $\eta = \tanh^2(\kappa L) = 0.996$. As the mismatch increases, the efficiency drops and shows an oscillatory behaviour. The first zero after the maximum in the efficiency curve occurs at a value of $\gamma = \pi/2L$, and thus the width W of the efficiency as function of the mismatch is given by:

$$W = \sqrt{\frac{\pi^2}{L^2} + \frac{\kappa^2}{4}} \tag{4.97}$$

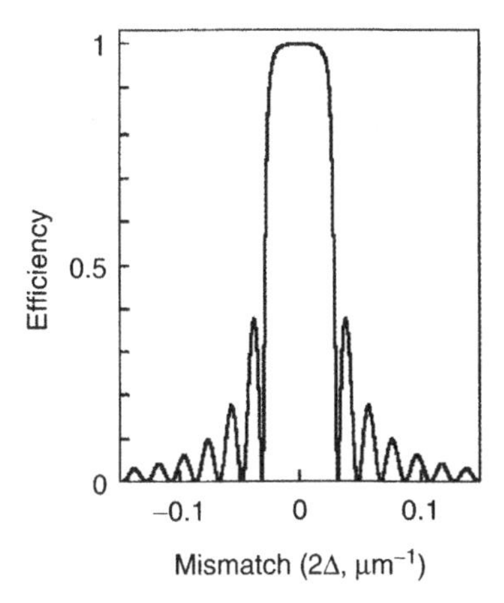

Figure 4.18 Efficiency of transfer between contrapropagating modes as a function of the mismatching parameter, for a coupling coefficient of $\kappa = 0.012\ \mu m^{-1}$ and a coupling length of $L = 300\ \mu m$

4.2 Diffraction Gratings in Waveguides

4.2.1 Waveguide diffraction gratings

Planar and channel waveguides with diffraction gratings are among the key components in the fabrication of integrated photonic devices [1, 2]. A diffraction grating, or grating for short, is a periodic structure with a period Λ comparable to the wavelength λ of the light being used. The main task of a grating integrated in a waveguide structure is to allow coupling between modes, that would otherwise remain independent. Under specific conditions, this characteristic permits the power transfer between collinear modes of different order, between contrapropagating modes, and also allows coupling between guided and radiation modes, and vice versa. Indeed, the use of integrated gratings has a great range of applications, because these structures have intrinsic selectivity in terms of wavelength, modal propagation constant of the modes involved in the coupling, or angle in the case of radiation modes. In addition, integrated photonics technology offers the possibility of easy and full integration of the periodic structures in planar or channel waveguides. One of the applications of waveguide gratings is their use as waveguide reflectors, that can be used to provide the necessary feedback in integrated lasers based on rare-earth doped materials [8] or in semiconductor lasers [9]. The advantage of using integrated gratings instead of attached mirrors is the narrow linewidth achieved on the laser emission and the high mechanical stability of the devices provided by the so-called distributed Bragg reflectors (DBRs); if the grating in the integrated semiconductor laser is placed along the gain region of the waveguide, the device is called a distributed feedback (DFB) laser. The high selectivity of the grating is also used for wavelength filtering applications, such as in dense wavelength demultiplexing. As we will show, the periodic structures in waveguides can be also used as mode converters, for applications such as power splitters. In addition, since the periodic structures can also induce coupling between guided and radiation modes, waveguide gratings are efficient components for performing functions such as light coupling and out-coupling the light outside the guide [10]. This method is of interest in cases where the direct end-coupling with the optical fibre is not possible or very difficult. Finally, one interesting application of waveguide gratings lies in the field of integrated sensors, that allow the fabrication of very compact, highly sensitive and selective devices [11]. In this case, the optical waveguide grating acts as a coupler sensor that responds to the change of optical refractive index of the liquid or gas medium and to the adsorption or binding of molecules onto the surface [12].

From the point of view of the geometry, there are two main types of waveguide integrated gratings. In the relief grating (or corrugated grating) the periodic modulation is achieved by a periodic change in the guiding film thickness, usually on the film–cover surface (Figure 4.19, top). The coupling coefficient induced by these structures is directly related to the thickness modulation $2h$, that has to be much lower than the film thickness in order to avoid high propagation losses. Relief gratings are static (fixed) structures, in the sense that they cannot be modified once the integrated photonic chip has been fabricated, usually by means of photolithographic and etching techniques.

The second kind of waveguide gratings is the index modulation type grating. In this case, the waveguide grating is "written" by inducing a periodic modulation in the refractive index of the guiding film (or adjacent regions). Figure 4.19 (bottom) shows the structure of a modulation index waveguide grating, where the index change has been induced in the whole guiding film. In this type of grating, the modulation can be permanent, by inducing the index change via UV illumination in photosensitive materials [13], or dynamic, where the modulation is achieved through acousto-optic or electro-optic effects by applying electric signals to periodic electrodes [14].

4.2.2 Mathematical description of waveguide gratings

Regardless of the type, a waveguide diffraction grating is a periodic structure that can be described by the spatial change on the dielectric permittivity $\Delta\varepsilon$ caused by the grating in the original waveguide structure, that is, in the canonical structure. As $\Delta\varepsilon$ is a periodic function, it can be expressed as a Fourier expansion of the form:

$$\Delta\varepsilon(x, y, z) = \sum_q \Delta\varepsilon_q(x) e^{-iq\mathbf{K}\mathbf{r}} \tag{4.98}$$

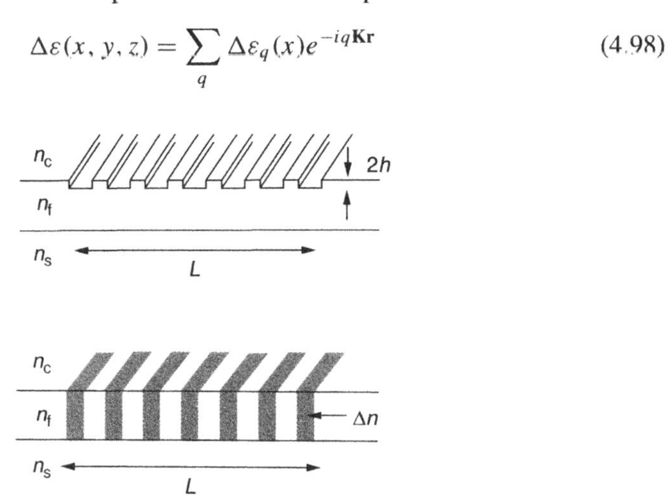

Figure 4.19 Two types of waveguide gratings. Above: relief grating. Below: index modulation grating

valid in the waveguide region where the grating is located, and $\Delta\varepsilon = 0$ outside that region. The quantities $\Delta\varepsilon_q(x)$ are the Fourier coefficients in the expansion, and they depend only on the x coordinate if the wavevector $\mathbf{K}$ associated with the grating lies in the plane defined by the optical waveguide (yz plane):

$$\mathbf{K} = K_y\mathbf{u}_y + K_z\mathbf{u}_z \tag{4.99}$$

$$\mathbf{r} = y\mathbf{u}_y + z\mathbf{u}_z \tag{4.100}$$

In addition, the magnitude of the wavevector $\mathbf{K}$ is related to the grating period Λ by the following formula:

$$|\mathbf{K}| = K = 2\pi/\Lambda \tag{4.101}$$

If a wave characterised by its wavevector $\mathbf{k}$ is incident in the grating region, as a result of the periodic phase modulation $\Delta\varepsilon$ induced by the periodic perturbation, several diffracted waves are generated, having wavevectors given by $\mathbf{k} + q\mathbf{K}$, q being an integer number. The existence of coupling between two guided modes induced by the grating implies that the propagation constants β_a and β_b of the modes must fulfil the relation:

$$\beta_b = \beta_a + qK \tag{4.102}$$

where q indicates the *coupling order*. This relation is the *Bragg condition* for guided modes in waveguides, and corresponds to the phase matching condition examined in the previous sections.

In order to obtain the modal coupling equations involving power transfer induced by waveguide gratings, one should start by rewriting equation (4.32) using the change on the dielectric permittivity given in (4.98). By proceeding in this way, we obtain for the variation of the $a_\mu(z)$ coefficients the following equation:

$$\pm\frac{da_\mu(z)}{dz} = -i\sum_\nu\sum_q \kappa_{\mu\nu}^{(q)} a_\nu(z) e^{-i(\beta_\nu + qK - \beta_\mu)z} \tag{4.103}$$

where the coupling coefficients of qth order are given by:

$$\kappa_{\mu\nu}^{t(q)} = \frac{\omega}{4}\iint \mathbf{E}_{\mu t}^*(x, y)\Delta\varepsilon_q(x)e^{-iqK_y y}\mathbf{E}_{\nu t}(x, y)\,dx\,dy \tag{4.104}$$

$$\kappa_{\mu\nu}^{z(q)} = \frac{\omega}{4}\iint E_{\mu z}^*(x, y)\Delta\varepsilon_q(x)e^{-iqK_y y}E_{\nu z}(x, y)\,dx\,dy \tag{4.105}$$

From (4.98) it follows that $\Delta\varepsilon_{-q} = \Delta\varepsilon_q^*$, and therefore the coupling coefficients fulfil the relation $\kappa_{\nu\mu}^{(-q)} = \kappa_{\mu\nu}^{(q)*}$. Although the right-hand side of equation (4.103) includes a summation respect to the coupling order q, in general one can ignore all the coupling orders except that corresponding to the efficient coupling between the modes μ and ν near the phase matching condition, because all the other terms oscillate very quickly with z and thus do not contribute appreciably to the coupling. Also, it is worth noting that the summation to all the modes ν includes also the term $\mu = \nu$, called the *self-coupling term*. This term, that describes the slight change on the propagation constant of a particular mode induced by the grating, is usually very small, and can be ignored in most cases.

4.2.3 Collinear mode coupling induced by gratings

Let us now consider the case of coupling between two collinear propagation modes a and b, that propagate along a planar waveguide in which exists a diffraction grating with wavevector K that perturbs the original structure. If the propagation constants of the modes β_a and β_b fulfil the Bragg condition (phase matching condition) for a particular coupling order q, then efficient power exchange can occur between them. Let us remember that without the presence of the periodic perturbation, the modes a and b were independent, and power transfer between them was impossible because of the orthogonality relation. In principle, for a planar waveguide the modes a and b can be any arbitrary combination of TE and TM modes.

A diffraction grating integrated in a planar waveguide can be mathematically expressed as:

$$\Delta\varepsilon(x,z) = \sum_q \Delta\varepsilon_q(x)e^{-iqKz} \tag{4.106}$$

If the propagation constants of the guided modes β_a and β_b fulfil, at least approximately, the phase matching condition for the qth coupling order ($\beta_b \approx \beta_a + qK$), and no other combination of modes exists that fulfils that relation, the modal coupling equations are reduced to:

$$\pm\frac{dA(z)}{dz} = -i\kappa^* B(z)e^{-i2\Delta z} \tag{4.107}$$

$$\pm\frac{dB(z)}{dz} = -i\kappa A(z)e^{+i2\Delta z} \tag{4.108}$$

where the self-coupling term has been ignored, and the coupling coefficient $\kappa_{ba}^{(q)}$ is denoted simply by κ. The *mismatching parameter* 2Δ is expressed in this case by:

$$2\Delta = \beta_b - (\beta_a + qK) \tag{4.109}$$

and denotes the deviation with respect to the perfect phase matching condition, taking into account the effect of the periodic structure via the qth order coupling (Figure 4.20). Besides the expression of the mismatching, the differential coupled equations (4.107) and (4.108) are similar to those found in the previous section, and thus their solutions are the solutions already described when we treated the general cases of collinear coupling.

Co-directional coupling Let us now examine the coupling between two guided modes a and b, induced by a periodic structure, that propagate in the same direction but having different propagation constants ($\beta_a \neq \beta_b$). If we assume propagation

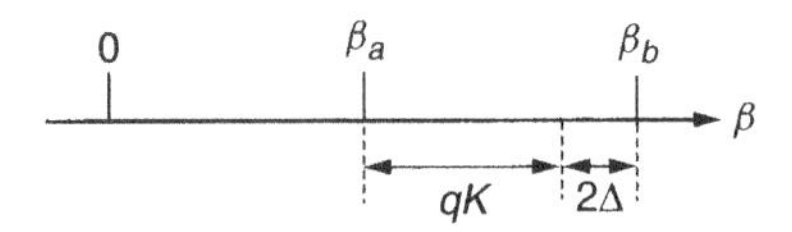

Figure 4.20 Relation between the propagation constants of the modes (β_a and β_b), the wavevector of the grating (K) and the mismatching parameter (2Δ) in the coupling induced by a waveguide grating structure

along the $+z$ direction (β_a, $\beta_b > 0$), then in equations (4.107) and (4.108) the plus sign in the left-hand side must be chosen. The solutions of these equations in the case of light injection corresponding to the selective excitation of mode a at $z = 0$ ($A(0) = 1$, $B(0) = 0$) are given by:

$$A(z) = e^{-i\Delta z}\left[\cos\gamma z + \frac{i\Delta}{\gamma}\sin\gamma z\right] \tag{4.110}$$

$$B(z) = e^{+i\Delta z}\frac{-i\kappa}{\gamma}\sin\gamma z \tag{4.111}$$

where the parameter γ is defined by:

$$\gamma = \sqrt{|\kappa|^2 + \Delta^2} \tag{4.112}$$

The power fluxes of the modes are given by the squared modulus of the $A(z)$ and $B(z)$ coefficients, and it yields:

$$\frac{|A(z)|^2}{|A(0)|^2} = 1 - F\sin^2\gamma z \tag{4.113}$$

$$\frac{|B(z)|^2}{|A(0)|^2} = F\sin^2\gamma z \tag{4.114}$$

where we have introduced for the sake of clarity the a-dimensional parameter F defined as:

$$F \equiv \frac{|\kappa|^2}{\gamma^2} = \frac{1}{1 + \Delta^2/|\kappa|^2} \tag{4.115}$$

From the above expressions it is clear that, as happened with the co-directional coupling between the modes in two coupled waveguides, the coupling between two modes in a waveguide induced by the presence of a periodic structure gives rise to a periodic transfer of power as the wave advances along the $+z$ direction (in the region where the periodic structure exists). The efficiency of power transfer between the modes a and b for a periodic structure of length L is given by:

$$\eta = \frac{|B(L)|^2}{|A(0)|^2} = \frac{\sin^2\gamma L}{1 + \Delta^2/|\kappa|^2} \tag{4.116}$$

In the situation of perfect phase matching condition ($\Delta = 0$), the efficiency simplifies to:

$$\eta = \eta_0 = \sin^2(|\kappa|L) \tag{4.117}$$

This last formula implies complete power transfer from the mode a to the mode b when the periodic structure has a length L given by the *coupling length* defined as: $L_c = \pi/(2|\kappa|)$.

Contradirectional coupling In this case the coupling takes place between two collinear modes a and b that propagate in opposite directions, that is, the propagation constants of the modes have opposite signs, for instance $\beta_a > 0$ and $\beta_b < 0$ (Figure 4.21). This kind of interaction can occurs also when the a and b modes are the same mode (in the same waveguide), but with opposite propagation directions.

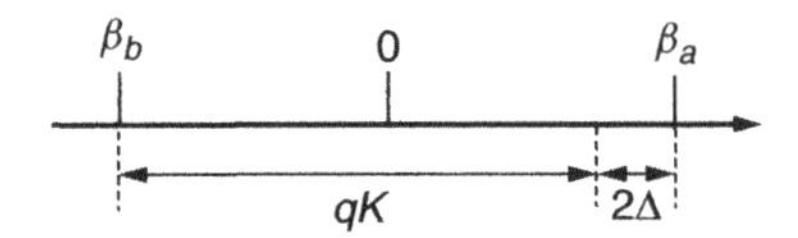

Figure 4.21 Scheme showing the coupling between contrapropagating modes induced by a waveguide grating

In the equations that describe the modal coupling one must choose the positive sign for equation (4.107) and the negative sign for (4.108). If we solve these equations in the case of light injection at $z = 0$ by exciting selectively the mode a, the initial boundary conditions are $A(0) = 1$ and $B(L) = 0$. The results, following the procedure given in the preceding section (and for a situation near the phase matching condition), are:

$$A(z) = e^{-i\Delta z} \frac{\alpha \cosh[\alpha(z - L)] + i\Delta \sinh[\alpha(z - L)]}{\alpha \cosh(\alpha L) - i\Delta \sinh(\alpha L)} \tag{4.118}$$

$$B(z) = e^{+i\Delta z} \frac{i\kappa \sinh[\alpha(z - L)]}{\alpha \cosh(\alpha L) - i\Delta \sinh(\alpha L)} \tag{4.119}$$

where the parameter α is defined by: $\alpha \equiv (|\kappa|^2 - \Delta^2)^{1/2} = -i\gamma$. The power fluxes as function of the position along the periodic structure (from $z = 0$ to $z = L$) are given by:

$$\frac{|A(z)|^2}{|A(0)|^2} = \frac{1 + F \sinh^2[\alpha(z - L)]}{1 + F \sinh^2(\alpha L)} \tag{4.120}$$

$$\frac{|B(z)|^2}{|A(0)|^2} = \frac{F \sinh^2[\alpha(z - L)]}{1 + F \sinh^2(\alpha L)} \tag{4.121}$$

where for clarity on the formulae we have introduced the parameter F defined as:

$$F \equiv \frac{|\kappa|^2}{\alpha^2} = \frac{1}{1 - \Delta^2/|\kappa|^2} \tag{4.122}$$

The equations (4.120) and (4.121), which describe the power evolution associated to the modes a and b, show a monotonic behaviour, in contrast to the periodic behaviour found in the case of co-directional coupling. The coupling efficiency, defined as the quotient of the power associated with the mode b at position $z = 0$ with respect to the power of the mode a at the same position, yields:

$$\eta = \frac{|B(0)|^2}{|A(0)|^2} = \left[\frac{\sinh^2 \alpha L}{\cosh^2 \alpha L - \Delta^2/|\kappa|^2}\right] \tag{4.123}$$

This formula indicates that total power transfer can only occur for an infinitely long periodic region. For a given periodic region length, the efficiency is maximum at the phase matching condition ($\Delta = 0$), where $\gamma = \kappa$ and $F = 1$, and in this case we have:

$$\eta = \frac{|B(0)|^2}{|A(0)|^2} = \tanh^2(|\kappa|L) \tag{4.124}$$

We showed that for a length of $L = \pi/|\kappa|$ the efficiency is greater that 0.99, indicating a nearly total transfer between the contrapropagating modes.

4.2.4 Coupling coefficients calculation

The coupling coefficients induced by periodic structures are the key magnitudes needed to calculate important parameters such as the optimum coupling length and the coupling efficiency. The calculation of these coefficients is performed by substituting the modal field profiles corresponding to modes a and b into the formulae given in (4.104)–(4.105). Depending on the polarisation character of the modes involved in the coupling (TE–TE, TE–TM, TM–TE or TM–TM), coupling coefficients can have very different values. Also, the geometry or type of periodic structure perturbation determines which of the various types of modes that a waveguide can support that interact each other efficiently. Here we discuss the calculation of the coupling coefficients in planar waveguides, which support pure TE or TM modes [15]. The results can be extrapolated to channel waveguides, providing that one of the lateral dimension is much greater that the other [16].

TE–TE coupling When the mutual coupled modes a and b are TE polarised modes in a planar waveguide, denoted by TE_μ and TE_ν respectively, the z-component of the electric field associated with these modes vanishes (see Chapter 3). This implies that the longitudinal coupling coefficient which involves E_z, is $\kappa^z_{TE_\mu TE_\nu} = 0$. Therefore, the coupling coefficient for TE–TE modes is the transversal coupling coefficient κ^t, given by

$$\kappa^{(q)}_{TE_\mu TE_\nu} = \kappa^{t(q)}_{TE_\mu TE_\nu} = \frac{\omega}{4} \int E^*_{\mu y}(x) \Delta\varepsilon_q(x) E_{\nu y}(x)\, dx \tag{4.125}$$

where E_y is the only non-vanishing transversal component of the electric field for TE modes. Let us remember that the electric field profiles into equation (4.125) have been normalised to the power flux.

TM–TM coupling In this case, the electric field associated with TM polarised modes has transversal as well as longitudinal components, and so in general two types of coupling coefficients must be taken into account, one of them associated with the transversal field component, and the other related to the longitudinal component of the electric field:

$$\kappa_{TM_\mu TM_\nu} = \kappa^t_{TM_\mu TM_\nu} + \kappa^z_{TM_\mu TM_\nu} \tag{4.126}$$

Nevertheless, as we showed in the section (4.1.3), in most of the cases the transversal component of the electric field is much greater that its longitudinal component, and the longitudinal coupling coefficient can be ignored in a first approximation. In this way, the coupling coefficient for TM–TM coupling can be approximated to:

$$\kappa^{(q)}_{\mu\nu} \approx \kappa^{t(q)}_{\mu\nu} = \frac{\beta_\nu \beta_\mu}{4\omega} \int \frac{H^*_{\mu y}(x)}{\varepsilon(x)} \Delta\varepsilon_q(x) \frac{H_{\nu y}(x)}{\varepsilon(x)} dx \tag{4.127}$$

where we have used the relation $E_x = (\beta/\omega\varepsilon)H_y$ for planar waveguides given by equation (3.18).

TE–TM coupling Let us now consider the case in which the two modes involved in the mutual coupling have orthogonal polarisation, that is, one mode is TE polarised and the second one is TM polarised. In principle, if the periodic structure is isotropic, as the TE_μ and TM_ν modes are orthogonal, no power transfer can exist between, and therefore the coupling coefficient is zero: $\kappa_{TE_\mu TM_\nu} = 0$. Nevertheless, if the periodic structure that perturbs the waveguide is anisotropic, under specific conditions, the coupling between orthogonal TE and TM modes can occur. This situation can be achieved in some anisotropic material through the electro-optic, acousto-optic or magneto-optic effects [17]. In particular, lithium niobate crystals and polymers exhibit a high value of the electrooptic effect, and can be used to fabricate a TE–TM converter via the TE–TM coupling [18, 19].

As we have shown in section (4.1.3), in the case of coupling induced by anisotropic changes in the permittivity tensor, coupling between mutual orthogonal polarisation modes can take place if the off-diagonal elements of the tensor are non-zero. For the particular case in which the change on the dielectric permittivity is of the form:

$$\Delta\varepsilon = \begin{pmatrix} \Delta\varepsilon_{xx} & \Delta\varepsilon_{xy} & 0 \\ \Delta\varepsilon_{yx} & \Delta\varepsilon_{yy} & 0 \\ 0 & 0 & \Delta\varepsilon_{zz} \end{pmatrix} \tag{4.128}$$

following the formula given in (4.45), the coupling coefficient between the TE_μ mode and the TM_ν mode is expressed as:

$$\kappa^{(q)}_{TE_\mu,TM_\nu} = \frac{\beta_\nu}{4}\int E^*_{\mu y}(x)\Delta\varepsilon^{(q)}_{yx}(x)\frac{H_{\nu y}(x)}{\varepsilon_{xx}(x)}\,dx \tag{4.129}$$

4.2.5 Coupling coefficients in modulation index gratings

A modulation index grating is a structure where the refractive index is periodically varied. In a planar optical waveguide a modulation index grating is generated by inducing a periodic change in the refractive index of the guiding film, or in a region close to it, without changing the waveguide geometry. This periodic change of the refractive index can be achieved by means of the acousto-optic effect (dynamic grating) via a piezo-electric transducer, or by UV illumination (permanent or static grating) using a phase mask or an interference light pattern.

If the change in the refractive index is uniform in the guiding film of the planar waveguide and perpendicular to the propagation direction (see Figure 4.22), it can be expressed mathematically as:

$$\Delta n(x,z) = \sum \Delta n_q(x)\cos(qKz + \Phi_q) \tag{4.130}$$

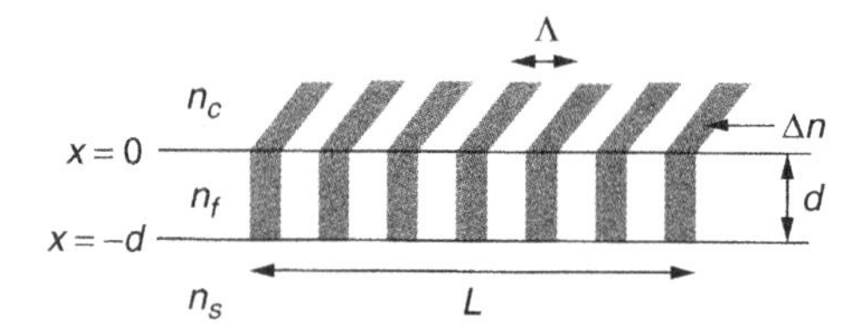

Figure 4.22 Modulation index grating geometry in a step-index planar waveguide

where $\Delta n_q(x) = \Delta n_q$ for $-d < x < 0$, and $\Delta n_q(x) = 0$ out of this region. If the refractive index change induced in the guiding film is small ($\Delta n \ll n$), the following approximation holds:

$$\varepsilon + \Delta\varepsilon = \varepsilon_0(n + \Delta n)^2 \approx \varepsilon_0(n^2 + 2n\Delta n) \Rightarrow \Delta\varepsilon = 2\varepsilon_0 n \Delta n \tag{4.131}$$

and thus the expansion terms in equation (4.106) can be expressed as:

$$\Delta\varepsilon_q(x) = \Delta\varepsilon_q = \varepsilon_0 n_f \Delta n_q e^{-i\Phi_q} \tag{4.132}$$

in the region $-d < x < 0$.

Taking into account the modal normalisation applied to planar waveguide structures for TE and TM modes (setting the power flow to 1 watt per unit width in the y-direction):

$$\frac{\beta}{2\omega\mu_0}\int_{-\infty}^{\infty} |E_{\mu y}|^2\, dx = 1 \qquad \text{TE modes} \tag{4.133}$$

$$\frac{\beta}{2\omega}\int_{-\infty}^{\infty} \frac{|H_{\mu y}|^2}{\varepsilon(x)}\, dx = 1 \qquad \text{TM modes} \tag{4.134}$$

the coupling coefficients for TE–TE and TM–TM modes yield:

$$\kappa_{TE_\mu,TE_\nu} \approx \frac{\pi\Delta n_q}{\lambda}\frac{n_f\displaystyle\int_{-d}^{0} E_{\mu y}^* E_{\nu y}\, dx}{\sqrt{N_\mu N_\nu}\left[\displaystyle\int_{-\infty}^{\infty} |E_{\mu y}|^2\, dx \int_{-\infty}^{\infty} |E_{\nu y}|^2\, dx\right]^{1/2}} \tag{4.135}$$

$$\kappa_{TM_\mu,TM_\nu} \approx \frac{\pi\Delta n_q}{\lambda}\frac{n_f\sqrt{N_\mu N_\nu}\displaystyle\int_{-d}^{0} (H_{\mu y}^*/\varepsilon)(H_{\nu y})/\varepsilon)\, dx}{\left[\displaystyle\int_{-\infty}^{\infty} (|H_{\mu y}|^2/\varepsilon)\, dx \int_{-\infty}^{\infty} (|H_{\nu y}|^2/\varepsilon)\, dx\right]^{1/2}} \tag{4.136}$$

where λ is the wavelength of light, n_f is the guiding film refractive index, and N denotes the effective refractive index of the mode.

The formula (4.135) can be simplified for modes far from cut-off. In this situation, the modes are very well confined within the guiding film ($-d < x < 0$), and the integral limits that appear in the numerator can be extended with good approximation to $\pm\infty$, thus:

$$\int_{-d}^{0} E_{\mu y}^* E_{\nu y}\, dx \approx \int_{-\infty}^{\infty} E_{\mu y}^* E_{\nu y}\, dx \approx \frac{2\omega\mu_0}{\beta}\delta_{\mu\nu} \tag{4.137}$$

where we have used the modal orthogonality, indicating that the second factor in (4.135) is close to one for modes with same order ($\mu = \nu$), and close to zero for different mode orders ($\mu \neq \nu$); in addition, the same reasoning can be applied to the formula given for TM modes. Thus, the coupling coefficients can be written as:

$$\kappa_{TE_\nu TE_\mu} \approx \kappa_{TM_\nu TM_\mu} \approx \frac{\pi\Delta n_q}{\lambda}\delta_{\mu\nu} \tag{4.138}$$

This last result indicates that the modal conversion between modes having different order ($\mu \neq \nu$) cannot be achieved by using uniform index modulation structures. On the contrary, this type of periodic structures is effective in producing strong interactions between modes of equal order ($\mu = \nu$), in particular for contradirectional coupling between a mode propagating in one direction and the same mode propagating in opposite direction. Therefore, the modulation index structures in waveguides are suitable for fabricating waveguide reflectors, in applications such as lasers and wavelength demultiplexors.

The factor $\pi \Delta n_q / \lambda$ included in the coupling coefficient for TE–TE and TM–TM conversion using modulation index gratings is in fact the coupling coefficient found for monochromatic plane waves in bulk. Thus, the formula given in (4.138) indicates that the coupling coefficient in this type of waveguide gratings is equal to that of contrapropagating plane waves. If the approximation of mode far from the cut-off does not hold, the coupling coefficient for waveguide modes differs from that calculated for plane waves. Figure 4.23 represents an example of this behaviour, where the ratio between the waveguide and bulk coupling coefficients ($\kappa_{wg}/\kappa_{bulk}$) has been plotted as a function of the film thickness, for a step-index planar waveguide where a modulation index grating has been uniformly induced in the guiding film. The effective refractive index of the fundamental TE mode has been also plotted in the same figure. For large film thickness, the effective refractive index of the TE_0 mode is very far from cut-off (close to the film index value), the mode energy is well confined within the film region, and thus the ratio $\kappa_{wg}/\kappa_{bulk}$ is very close to one. As the film thickness decreases, the mode effective refractive index N moves away from n_f, and the mode energy spreads out to the cover and substrate regions. Therefore the approximation assumed in equation (4.137) is no longer valid, and the ratio $\kappa_{wg}/\kappa_{bulk}$ decreases. When the mode index reaches values close to the cut-off value ($N \approx n_s$), the energy is not longer confined in the film and the coupling coefficient for the waveguide grating κ_{wg} drops to zero.

When modulation index gratings are used as narrow-band wavelength filters, it is important that the reflectivity curve is very narrow as a function of the mismatch (or

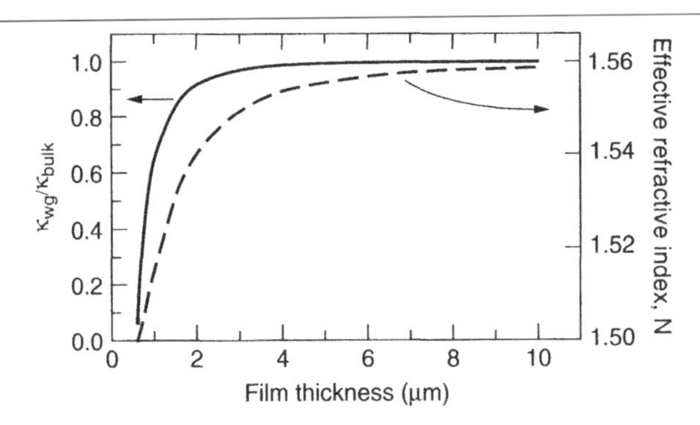

Figure 4.23 Ratio between bulk and waveguide coupling coefficients. The waveguide coupling coefficient corresponds to TE_0–TE_0 contrapropagating guided modes in a step-index planar waveguide, consisting of a SiO_2/GeO_2 film ($n_f = 1.56$) on a SiO_2 substrate ($n_s = 1.50$), at a working wavelength of $\lambda = 1.3$ μm. The refractive index of the guiding film has been modulated by UV radiation. The cover is assumed to be air ($n_c = 1.00$)

the wavelength, which is closely related to the mismatch through the mode propagation constants). Following the formula given in (4.97), this can be done by fabricating long gratings with small coupling coefficients (that is, small modulation index changes). Nevertheless, the sidelobes that appear in the reflectivity curve induce spurious reflectivity at other wavelengths apart from the central peak wavelength. One way of avoiding this problem is to suppress the sidelobes, which can be achieved by designing an *apodised grating*. In such apodised waveguide grating the modulation depth in the refractive index is changed along the z direction of the structure. One common functional shape of the apodisation is given by:

$$\Delta n = \Delta n_0 \sin^2(\pi z/L) \tag{4.139}$$

where the Δn_0 is the maximum modulation depth and L is the grating length. The reflectivity characteristics of such non-uniform gratings can be calculated following the algorithm developed in [20], which also includes the possibility of considering absorption or gain in the grating region. Figure 4.24 shows the reflectivity behaviour of a uniform modulation index grating compared to an apodised grating using the same coupling and grating length parameters, where the suppression of the sidelobes is apparent.

4.2.6 Coupling coefficients in relief diffraction gratings

Relief diffraction gratings achieve the modulation on $\Delta\varepsilon$ by a creating a periodic change in the waveguide dimensions. If the modification consists of periodic step-like changes of the guiding film thickness of a planar waveguide, the diffraction grating will have a rectangular profile as shown in Figure 4.25. The Fourier terms describing this type of periodic structure are given by:

$$\Delta\varepsilon_q(x) = \Delta\varepsilon_q = \varepsilon_0(n_f^2 - n_c^2)\frac{\sin q a \pi}{q\pi} \tag{4.140}$$

with $q \neq 0$, $0 < a < 1$, in the region corresponding to $-h < x < h$, and $\Delta\varepsilon_q(x) = 0$ elsewhere.

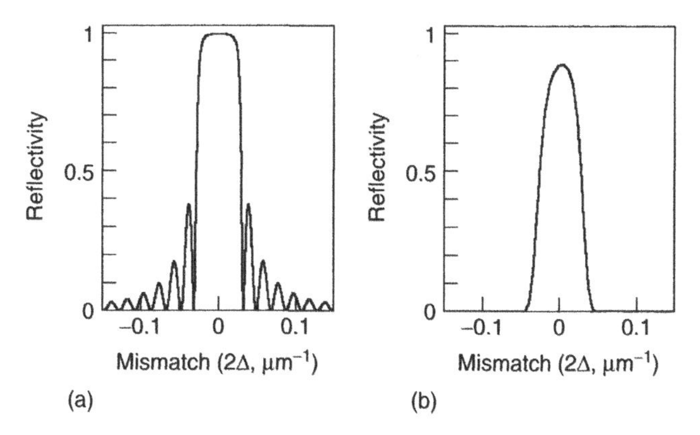

Figure 4.24 Reflectivity of a modulation index waveguide grating as function of the mismatch parameter. (a) Uniform grating. (b) Apodised grating

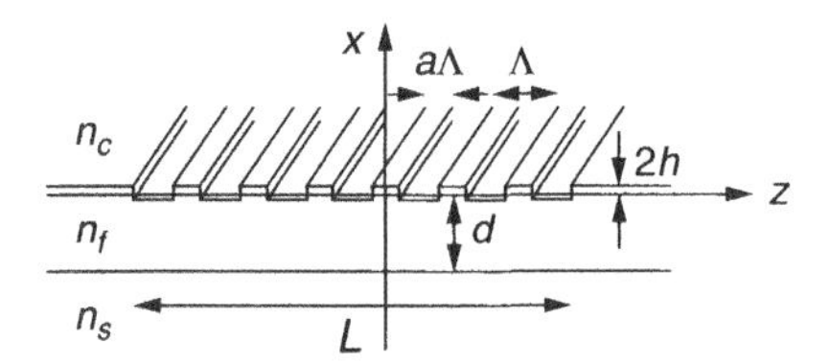

Figure 4.25 Geometry of a step-index planar waveguide with a rectangular relief diffraction grating

The coupling coefficients corresponding to this perturbed geometry are calculated by introducing (4.140) into equations (4.104)–(4.105), and evaluating the integral. In the case of TE–TE coupling one obtains:

$$\kappa^{(q)}_{TE_\mu TE_\nu} = \frac{\pi}{\lambda}\frac{(n_f^2 - n_c^2)}{\sqrt{N_\nu N_\mu}}\frac{\sin q\pi a}{q\pi}\frac{\displaystyle\int_{-h}^{h} E^*_{\mu y} E_{\nu y}\, dx}{\sqrt{\displaystyle\int_{-\infty}^{\infty} |E_{\nu y}|^2\, dx \int_{-\infty}^{\infty} |E_{\mu y}|^2\, dx}} \tag{4.141}$$

A similar expression for TM–TM coupling is found, but is now expressed as a function of the H_y component of the fields.

Figure 4.26 shows the coupling coefficients calculated using the formula (4.141) for an asymmetric planar waveguide, where a rectangular relief grating is situated at the film–cover interface, using order $q = 1$. We observe that coupling is allowed for modes having different order, and the coupling coefficient is higher for high order modes. This behaviour is explained by the fact that high order modes penetrate the relief grating more where the coupling takes place.

A good approximation of the coupling coefficients can be obtained if the modulation depth of the periodic structure is much lower than the waveguide film thickness, that is, when $2h \ll d$. In this case, the electric or magnetic field in the numerator in the

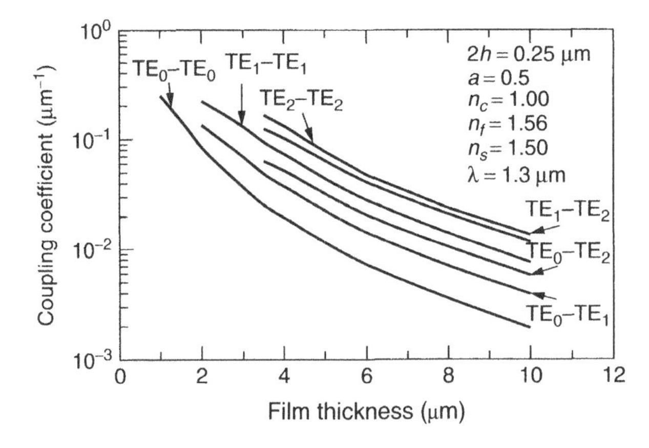

Figure 4.26 Coupling coefficients for TE_μ–TE_ν modes in a step-index planar waveguide as function of the guiding film thickness. Coupling is induced by a rectangular relief diffraction grating with thickness $2h$, situated at the cover-film interface, using the coupling order $q = 1$

integral of the coupling coefficient formulae can be considered constant, and its value is the electric or magnetic field evaluated at $x = 0$.

One problem found when working with rectangular relief gratings is that several diffraction orders are involved in the coupling, and it is therefore difficult to control and isolate the coupling between two particular modes. This is avoided if the relief diffraction grating has a sinusoidal profile. In this case, if the corrugation function is $h\cos(Kz)$, the change on the dielectric permittivity induced in the waveguide structure is expressed as:

$$\Delta\varepsilon(x,z) = \begin{cases} +(n_f^2 - n_c^2) & 0 < x < h\cos Kz \\ -(n_f^2 - n_c^2) & h\cos Kz < x < 0 \end{cases} \tag{4.142}$$

The coupling coefficient is once again calculated by using this expression in formulae (4.104) and (4.105), bearing in mind that now $\Delta\varepsilon$ depends not only on the x coordinate but also on the z coordinate. The result is that for coupling orders $|q| \neq 1$ the coupling coefficients between TE–TE and TM–TM modes are very small, thus the energy transfer via high order coupling do not take place. For the fundamental order $q = 1$ the coupling coefficient for TE–TE modes is given by [21]:

$$\kappa_{TE_\mu TE_\nu}^{(q)} = \frac{\pi h}{2\lambda}\frac{(n_f^2 - n_c^2)}{\sqrt{N_\nu N_\mu}}\frac{E_{\mu y}^*(0)E_{\nu y}(0)}{\sqrt{\int_{-\infty}^{\infty}|E_{\nu y}|^2\,dx\int_{-\infty}^{\infty}|E_{\mu y}|^2\,dx}} \tag{4.143}$$

where we have assumed that the electric field amplitude does not vary across the corrugation height ($h \ll d$), and $E(0)$ denotes the electric field value at the waveguide-cover interface.

As an illustration of mode coupling induced by sinusoidal relief grating, Figure 4.27 shows the intensity evolution (in grey scale) as a function of the propagation distance in a symmetric waveguide having a sinusoidal corrugated diffraction grating located at one of the film–substrate interfaces. After launching light corresponding to the fundamental mode, the grating induces mode conversion to the TE_2 mode. Indeed, the grating period has been calculated according to the relation given in (4.102), for a

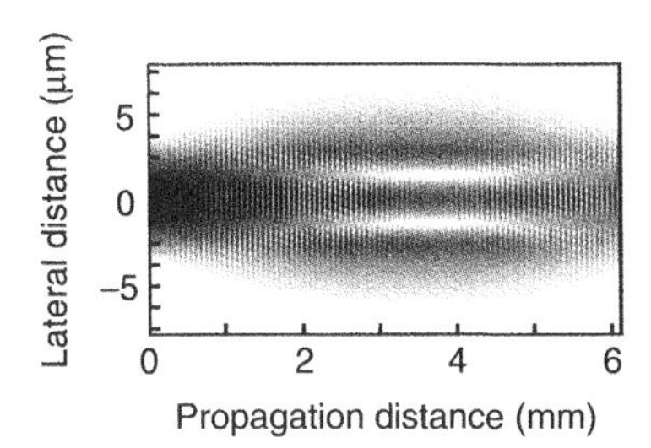

Figure 4.27 Light propagation in a symmetric step index planar waveguide, having a sinusoidal relief grating, after launching light into the fundamental mode. Waveguide parameters: $n_f = 2.203$, $n_c = n_s = 2.200$, $d = 5$ μm, $\lambda = 1.55$ μm. Parameters of the grating: $\Lambda = 67.0$ μm, $h = 0.1$ μm. After a propagation distance of $z \approx 3.9$ mm the fundamental mode is totally converted to the mode TE_2

perfect phase matching condition between the TE_0 and TE_2 modes, taking into account their propagation constants.

When the two modes are present, a beat appears in the intensity pattern, because the modes have different propagation constant. As the wave advances along the waveguide, the energy of the fundamental mode transfers to the TE_2 mode, and after a propagation distance of $z \approx 3.9$ mm all the initial power is fully transferred to the TE_2 mode (coupling length). If propagation continues, the energy transfers back to the fundamental TE_0 mode.

In summary, the coupling associated to $TE_\mu \leftrightarrow TE_\nu$ and $TM_\mu \leftrightarrow TM_\nu$ conversion with $\mu \neq \nu$ can be achieved by using relief index structures, in contrast to the situation faced when working with modulation index gratings. This difference comes from the fact that while in modulation index structures the modulation is uniform in the guiding film, in relief gratings the perturbation is located in a region close to the film surface.

References

[1] A. Yariv and M. Nakamura, "Periodic Structures for Integrated Optics", *IEEE Journal of Quantum Electronics*, **QE-13**, 233–253 (1977).

[2] T. Suhara and H. Nishihara, "Integrated Optics Components and Devices Using Periodic Structures", *IEEE Journal of Quantum Electronics*, **QE-22**, 845–867 (1986).

[3] D. Marcuse, "Coupled-Mode Theory for Anisotropic Optical Waveguides", *The Bell System Technical Journal*, **54**, 985–995 (1975).

[4] W.P. Huang, "Coupled-Mode Theory for Optical Waveguides: An Overview", *Journal of the Optical Society of America A*, **11**, 963–983 (1994).

[5] A. Yariv, "Coupled-Mode Theory for Guided-Wave Optics", *IEEE Journal of Quantum Electronics*, **QE-9**, 919–933 (1973).

[6] N. Nishihara, M. Haruna and T. Suhara, *Optical Integrated Circuits*, Mac-Graw Hill, R.R. Donelley & Sons Company, United States of America (1989).

[7] R. Ulrich, "Efficiency of Optical-Grating Couplers", *Journal of the Optical Society of America*, **63**, 1419–1431 (1973).

[8] J.E. Roman and K.A. Winick, "Neodymium-Doped Glass Channel Waveguide Laser Containing an Integrated Bragg Reflector", *Applied Physics Letters*, **61**, 2744–2748 (1992).

[9] W.P. Huang, Q. Guo and C. Wu, "A Polarization-Independent Distributed Bragg Reflector Based on Phase-Shifted Grating Structures", *IEEE Journal of Lightwave Technology*, **14**, 469–473 (1996).

[10] S. Ura and S.J. Sheard, "A Configuration for Guided-Wave Excitation into a Disposable Integrated-Optic Head", *Optics Communications*, **146**, 85–89 (1998).

[11] M. Wicki, R.E. Kunz, G. Voirin, K. Tiefenthaler and A. Bernard, "Novel Integrated Optical Sensor Based on a Grating Coupler Tript", *Biosensors and Bioelectronics*, **13**, 1181–1185 (1998).

[12] D. Clerc and W. Lukosz, "Direct Inmunosensing with an Integrated-optical Output Grating Coupler", *Sensors and Actuators*, **B40**, 53–58 (1997).

[13] J.E. Roman and K.A. Winick, "Photowritten Gratings in Ion-Exchanged Glass Waveguides". *Optics Letters*, **18**, 808–810 (1993).

[14] H. Herrmann, K. Schäfer and Ch. Schmidt, "Low Loss, Tunable Integrated Acousto-Optical Wavelength Filter in $LiNbO_3$ with Strong Sidelobe Suppression", *IEEE Photonics Technology Letters*, **10**, 120–123 (1998).

[15] H.A. Haus and R.V. Schmidt, "Approximate Analysis of Optical Waveguide Grating Coupling Coefficients", *Applied Optics*, **15**, 774–781 (1976).

[16] K.A. Winick, "Effective-Index Method and Coupled-Mode Theory for Almost-Periodic Waveguide Gratings: A Comparison", *Applied Optics*, **31**, 757–764 (1992).
[17] L.N. Binh and J. Livingstone, "A Wide-Band Acoustooptic TE-TM Mode Converter Using a Doubly Confined Structure", *IEEE Journal of Quantum Electronics*, **QE-16**, 964–971 (1980).
[18] H. Porte, J.P. Goedgebuer, R. Ferriere and N. Fort, "Integrated TE-TM Mode Converter on Y-cut Z-Propagating $LiNbO_3$ with an Electrooptic Phase Matching for Coherence Multiplexing", *IEEE Journal of Quantum Electronics*, **25**, 1760–1762 (1989).
[19] Y.H. Wol, J.K. Jang, Z. Taehyoung, C.O. Min and Y.S. Sang, "TE-TM Mode Converter in a Poled-Polymer Waveguide", *IEEE Journal of Quantum Electronics*, **32**, 1054–1062 (1996).
[20] H.V. Baghdasaryan and T.M. Knyazyan, "Modelling of Strongly Nonlinear Sinusoidal Bragg Gratings by the Method of Single Expression", *Optical and Quantum Electronics*, **32**, 869–883 (2000).
[21] H.A. Haus and R.V. Schmidt, "Approximate Analysis of Optical Waveguide Grating Coupling Coefficients", *Applied Optics*, **15**, 774–781 (1976).

Further Reading

Optical and Quantum Electronics, volume 32, number 6/8 (2000), special issue on "Optical Waveguide Theory and Numerical Modelling".
A.A. Barybin and V.A. Dmitriev, *Modern Electrodynamics and Coupled-Mode Theory: Application to Guided-Wave Optics*. Rinton Press, Princeton, New Jersey, USA 2002.
P. Paddon and J.F. Young, "Two-Dimensional Vector-Coupled-Mode Theory for Textured Planar Waveguides", *Physical Review B*, **61**, 2090–2101 (2000).

5

LIGHT PROPAGATION IN WAVEGUIDES: THE BEAM PROPAGATION METHOD

Introduction

One of the fundamental aspects in integrated optics is the analysis and simulation of electromagnetic wave propagation in photonics devices based on waveguide geometries, including optical waveguides.

The problem to be solved is the following: given an arbitrary distribution of refractive index $n(x, y, z)$, and for a given wave field distribution at the input plane at $z = 0$, $E(x, y, z = 0)$, the spatial distribution of light $E(x, y, z)$ at a generic point z must to be found.

Figure 5.1 outlines the problem applied to a beam splitter. In this case, the distribution of the refractive index is known, which defines the optical circuit. When a light beam is injected at $z = 0$, the problem is to determine the light intensity distribution at the exit, and in particular, what will be the output light intensity in each of the two branches of the splitter.

In this chapter we describe the "beam propagation method" (BPM) applied to the study of light propagation in integrated photonics devices based on optical waveguides. We begin by deriving a paraxial form of the Helmholtz equation, known as the Fresnel equation. This equation, valid for paraxial propagation in slowly varying optical structures, is the starting point to develop BPM algorithms. We will describe the BPM based on the fast Fourier transform method and the BPM algorithm based on finite differences. We will show that in general the latter has superior performance for simulating light propagation in integrated optical elements. Also we present the implementation of transparent boundary conditions and filtering techniques, which are efficient ways of avoiding errors in the simulation of light propagation due to the finite size of the computational window and the use of large propagation steps, respectively. Finally, a modal description based on BPM is presented, which allows not only calculation of the modes supported by a straight waveguide, but also determines the modal weight associated with each individual guided mode.

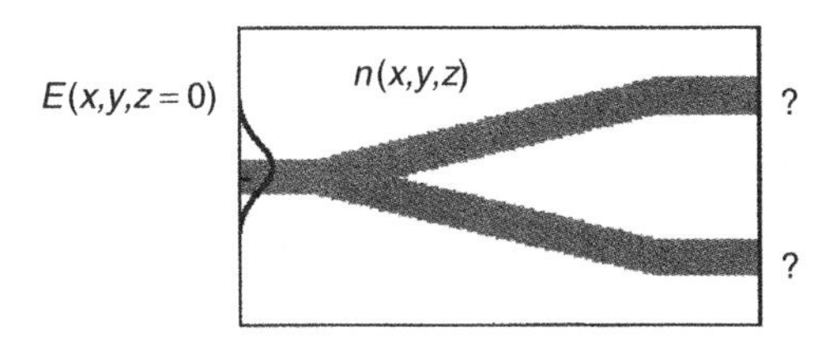

Figure 5.1 General problem in optical propagation: given an index distribution $n(x, y, z)$, and a wave field at the input, the light distribution at the exit must be found

5.1 Paraxial Propagation: Fresnel Equation

The problem of light propagation in waveguides with arbitrary geometry is very complicated in general, and it is necessary to make some approximations. In the first place, we will assume an harmonic dependence of the electric and magnetic fields, in the form of monochromatic waves with an angular frequency ω, in such a way that the temporal dependence will be of the form $e^{i\omega t}$. The equation which describes such EM waves is the vectorial Helmholtz equation:

$$\nabla^2 \mathbf{E} + \left(\frac{\omega}{c}\right)^2 n^2(x, y, z)\mathbf{E} = 0 \tag{5.1}$$

Although it is possible to work with this vectorial equation, in most cases it is possible to treat the optical propagation problem starting from the scalar Helmholtz equation. In this case the equation is of the form:

$$\nabla^2 E + \left(\frac{\omega}{c}\right)^2 n^2(x, y, z)E = 0 \tag{5.2}$$

where now $E = E(x, y, z)$ denotes each of the six Cartesian components of the electric and magnetic fields.

The refractive index in the domain of interest is given by $n(x, y, z)$, and will be determined by the waveguide geometry (optical fibre, directional coupler, Mach-Zehnder interferometer, etc.).

If the wave propagation is primarily along the positive z direction, and the refractive index changes slowly along this direction, the field $E(x, y, z)$ can be presented as a complex field amplitude $u(x, y, z)$ of slow variation, multiplied by a fast oscillating wave moving in the $+z$ direction (propagation direction):

$$E(x, y, z) = u(x, y, z)e^{-iKz} \tag{5.3}$$

where K is a constant which represents the characteristic propagation wave vector, $K = n_o\omega/c$, and n_0 is chosen, for example, as the refractive index of the substrate (or the cover).

Substituting the optical field in the Helmholtz equation, it follows that:

$$-\frac{\partial^2 u}{\partial z^2} + 2iK\frac{\partial u}{\partial z} = \left(\frac{\partial^2}{\partial x^2} + \frac{\partial^2}{\partial y^2}\right)u + (k^2 - K^2)u \tag{5.4}$$

where $k_0 = \omega/c = 2\pi/\lambda$ denotes the wavevector in the vacuum, and the notation $k(x, y, z) = k_0 n(x, y, z)$ has been introduced to represent the spatial dependence of the wavevector.

If we also assume that the optical variation is slow in the propagation direction ("slowly varying envelope approximation, SVEA"), we will have:

$$\left|\frac{\partial^2 u}{\partial z^2}\right| \ll \left|2K\frac{\partial u}{\partial z}\right| \tag{5.5}$$

In this case we can ignore the first term on the left-hand side of equation (5.4) with respect to the second one; this approximation is known as *parabolic* or *Fresnel approximation*, and equation (5.4) leads to:

$$2iK\frac{\partial u}{\partial z} = \left(\frac{\partial^2}{\partial x^2} + \frac{\partial^2}{\partial y^2}\right)u + (k^2 - K^2)u \tag{5.6}$$

which is known as the *Fresnel* or *paraxial equation.* It is the starting equation for the description of optical propagation in inhomogeneous media, and in particular, in waveguide structures. An example is TE propagation in 1D waveguides, where the Fresnel equation reduces to:

$$2ik_0 n_0\frac{\partial E_y}{\partial z} = \frac{\partial^2 E_y}{\partial x^2} + k_0^2[n^2(x, z) - n_0^2]E_y \tag{5.7}$$

where E_y is the only non-vanishing component of the electric field associated to TE modes of the 1D waveguide, and where the refractive index is represented by $n(x, z)$.

The solution to the Helmholtz equation or the Fresnel equation applied to optical propagation in waveguides is known as the *beam propagation method* (BPM). Two numerical schemes have been proposed to solve the Fresnel equation. In one of them, optical propagation is modelled as a plane wave spectrum in the spatial frequency domain, and the effect of the medium inhomogeneity is interpreted as a correction of the phase in the spatial domain at each propagation step [1]. The use of the *fast Fourier techniques* connects the spatial and spectral domains, and this method is therefore called *fast Fourier transform BPM* (FFT-BPM). The propagation of EM waves in inhomogeneous media can also be described directly in the spatial domain by a *finite difference scheme* (FD) [2]. This technique allows the simulation of strong guiding structures, and also of structures that vary in the propagation direction. The beam propagation method which solves the paraxial form of the scalar wave equation in an inhomogeneous medium using the finite difference method is called FD-BPM. Also methods based on finite differences which solve the vectorial wave equation, called FD-VPBM, have been developed [3]. There is an intermediate approximation, which starts from the wave equation but ignores coupling terms between the transversal components of the fields, and for that reason this method is usually referred to as *semi-vectorial* (FD-SVBPM).

5.2 Fast Fourier Transform Method (FFT-BPM)

The solution to the Helmholtz equation in a homogeneous medium characterised by a refractive index n_0 is a set of plane waves, and therefore the general solution can be represented by a superposition of such plane waves:

$$E(x, y, z) = \int_{-\infty}^{\infty} \int_{-\infty}^{\infty} E(k_x, k_y) e^{-ik_x x} e^{-ik_y y} e^{-ik_z z} dk_x dk_y \tag{5.8}$$

where:

$$k_z = [K^2 - k_x^2 - k_y^2]^{1/2} \tag{5.9}$$

and $K = n_0 k_0$. The amplitudes $E(k_x, k_y)$ can be obtained from the electric field distribution at $z = 0$ taking the Fourier transform of equation (5.8):

$$E(k_x, k_y) = \frac{1}{2\pi} \int_{-\infty}^{\infty} \int_{-\infty}^{\infty} E(x, y, 0) e^{ik_x x} e^{ik_y y} dx dy = F[E(x, y, 0)] \tag{5.10}$$

where F represents the Fourier transform operation. The algorithm for calculating the field at an arbitrary plane perpendicular to the z axis can be obtained by combining equation (5.8) with (5.10):

$$E(x, y, z) = F^{-1}\{F[E(x, y, 0)] e^{-ik_z z}\} \tag{5.11}$$

where F^{-1} represents the inverse Fourier transform. This is the *field diffractor operator* which represents the propagation in a medium characterised by the reference refractive index n_0.

The effect of the index variation $n(x, y, z)$ when the wave propagates a distance Δz is a small perturbation of the phase of the distribution of the phase front. This effect can be described by multiplying the field by a *lens corrector operator*, defined by $e^{-ik_0 \Delta n^2 \Delta z/2n_0}$, where $\Delta n^2 = n^2(x, y, z) - n_0^2$.

The BPM algorithm for the propagation along an arbitrary distance z is realised through several discrete steps of distance Δz, using a combination of the lens and diffraction operators (Figure 5.2), as indicated by the following operator sequence:

$$E(x, y, z + \Delta z) = F^{-1}\{e^{-ik_z \Delta z/2} F\{e^{-iK\Delta n^2 \Delta z/2n_0} F^{-1}[e^{-ik_z \Delta z/2} F[E(x, y, z)]]\}\} \tag{5.12}$$

5.2.1 Solution based on discrete fourier transform

The former method can be implemented numerically using discrete Fourier transform. In this case, the spatial domain of interest, of dimensions LxL in the transversal

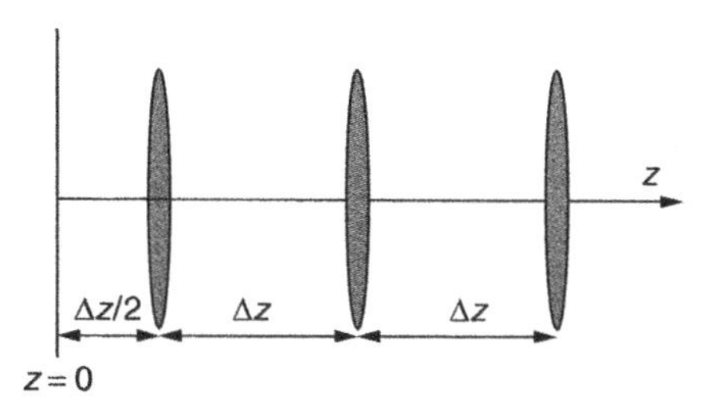

Figure 5.2 The algorithm FFT-BPM for solving the Helmholtz equation replaces the waveguide structure for a lens system. Between the lenses the electric field satisfies the Helmholtz equation in a homogeneous medium

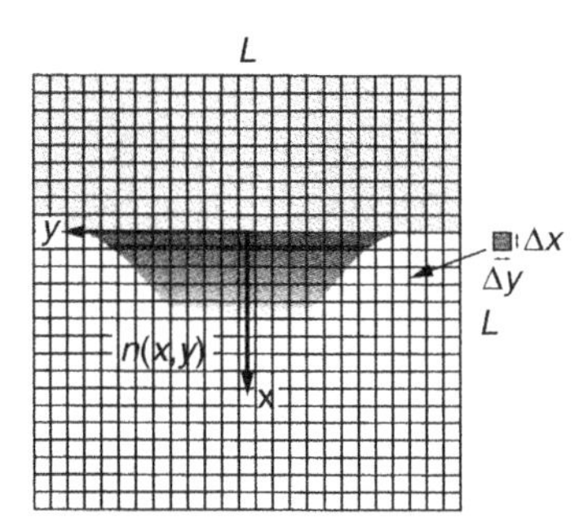

Figure 5.3 Computational window over the spatial domain of interest, that must include the waveguide structure, as well as the evanescent waves of the propagation modes

direction, is divided in a mesh of N^2 discrete points. This region must include not only the waveguide structure, but also must be wide enough to describe at least the evanescent waves corresponding to the propagation modes. Each cell has dimensions of $\Delta x \Delta y$, being $\Delta x = \Delta y = L/N$ (Figure 5.3).

In the optical propagation in a homogeneous medium having a constant refractive index n_0, the solution for the electric field given by the equation (5.8) can be described through a finite series in terms of the spatial frequencies:

$$E_{j,m}(z) = \sum_{\mu=-N/2+1}^{N/2} \sum_{\nu=-N/2+1}^{N/2} E_{\mu\nu}(0) e^{-i\frac{2\pi}{N}(j\mu+m\nu)} e^{-ik_z z} \tag{5.13}$$

where the spatial coordinates x and y are expressed as:

$$x = j\Delta x = jL/N, \quad y = m\Delta y = mL/N \tag{5.14}$$

and the discrete spatial frequencies k_x and k_y are expressed as:

$$k_x = (2\pi/L)\mu, \quad k_y = (2\pi/L)\nu \tag{5.15}$$

The description of equation (5.10) using discrete Fourier transforms is expressed as:

$$E_{\mu\nu}(0) = \frac{1}{N^2} \sum_{j=1}^{N} \sum_{m=1}^{N} E_{jm}(0) e^{i\frac{2\pi}{N}(j\mu+m\nu)} \tag{5.16}$$

The propagation through a step Δz starts with a propagation along the homogeneous medium to a distance $\Delta z/2$ (Figure 5.2), that can be calculated using the following discrete expansion:

$$E_{jm}(z + \Delta z/2) = \sum_{\mu=-N/2+1}^{N/2} \sum_{\nu=-N/2+1}^{N/2} E_{\mu\nu}(z) e^{-i\frac{2\pi}{N}(j\mu+m\nu)} e^{-ik_z \Delta z/2} \tag{5.17}$$

The next operation is to calculate the spatial frequency spectrum of the field, having in mind the perturbation induced by the phase change due to the medium inhomogeneity, because $n = n(x, y, z)$. The lens corrector operator acting through the electric field takes account of that perturbation, and therefore the series expansion in this case will be:

$$E_{\mu\nu}(z+\Delta z) = \frac{1}{N^2}\sum_{j=1}^{N}\cdot\sum_{m=1}^{N} E_{jm}(z+\Delta z/2)e^{-ik_0\Delta n^2\Delta z/2n_0}e^{i\frac{2\pi}{N}(j\mu+m\nu)} \qquad (5.18)$$

Finally, the propagation through the discrete step Δz ends with a propagation along a distance $\Delta z/2$ in a homogeneous medium. This step is described by:

$$E_{jm}(z+\Delta z) = \sum_{\mu=-N/2+1}^{N/2}\ \sum_{\nu=-N/2+1}^{N/2} E_{\mu\nu}(z+\Delta z)e^{-i\frac{2\pi}{N}(j\mu+m\nu)}e^{-ik_z\Delta z/2} \qquad (5.19)$$

With this sequence it is possible to calculate the electric field at any arbitrary plane perpendicular to the z axis, combining steps of length Δz.

Although the direct implementation of expressions (5.16)–(5.19) involves a number of calculations proportional to N^3, the use of fast Fourier transformation techniques allows us to reduce the number of calculations which scale as $N^2\, log_2 N$. If the calculation is performed in a single dimension (in the case of planar waveguides), the number of operations is reduced to $N\, log_2 N$ using FFT [4].

Figure 5.4 shows an example of light propagation in a symmetric planar waveguide using FFT-BPM algorithm. The waveguide has been designed to be monomode at the working wavelength of 1 μm. As the input optical field does not correspond exactly to the guided mode profile, the optical field smoothly changes along the propagation trying to accommodate the profile of the real propagation mode. The detection of noise in the field profile reveals the fact that the chosen propagation step of 2 μm is an upper limit for properly simulating such a structure using FFT-BPM. This problem becomes more important when 2D structures must be modelled, and although it is possible to implement FFT-BPM to simulate light propagation in such structures, another numerical technique is preferred, based on finite difference schemes.

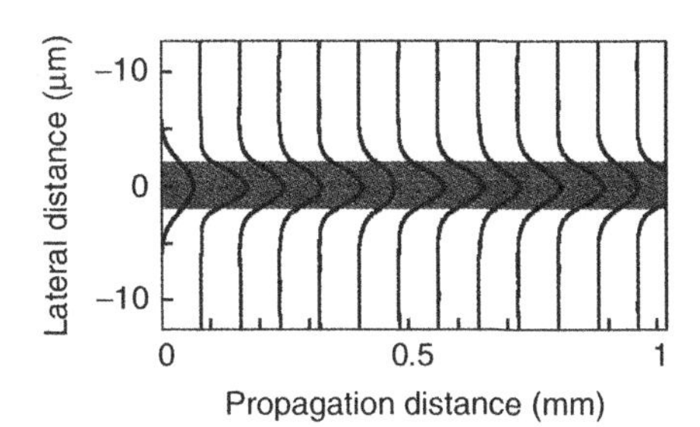

Figure 5.4 Light propagation in a one-dimensional waveguide monomode at $\lambda = 1$ μm, following a FFT-BPM scheme. The structure consists of a symmetric step-index planar waveguide, uniform in its propagation direction, with the following parameters: $n_0 = 2.2$, $\Delta n = 0.003$, $d = 4$ μm. The numerical simulation has been carried out with the parameters: mesh of 256 points, with $\Delta x = 0.1$ μm ($L = 25.6$ μm); propagation step $\Delta z = 2$ μm, over a distance of 1024 μm. The input optical field is a 4 μm-wide Gaussian distribution

5.3 Method Based on Finite Differences (FD-BPM)

The numerical method based on finite differences described in this section will be developed for one-dimensional structures (planar waveguides). In order to solve optical propagation in 2D waveguides (channel waveguides, optical fibres, etc.) using finite differences methods, one usually reduces the 2D-problem to a one-dimensional problem through the effective index method described in Chapter 3. This procedure gives very good results in general, and in addition saves computer time and releases a lot of computer memory. Nevertheless, it is also possible to develop finite difference schemes for 2D-structures [3].

In the FD-BPM method, the Helmholtz scalar wave equation (equation (5.6)) in partial derivatives is approximated by a finite difference scheme, which can be expressed as:

$$2iK\frac{u_j(z+\Delta z)-u_j(z)}{\Delta z}=\frac{u_{j-1}(z)-2u_j(z)+u_{j+1}(z)}{\Delta x^2}+k_0^2(n^2-n_0^2)u_j(z) \quad (5.20)$$

where $u_j(z)$ is the optical filed at the position $(j\Delta x, z)$ with $j=1,2,\ldots,N$. This scheme in finite differences, known as "*forward-difference*", allows us to calculate the optical field $u_j(z+\Delta z)$ after a propagation step Δz from a knowledge of the complete field $u_j(z)$ at the position z [5]. The calculation of $u_j(z+\Delta z)$ from equation (5.20) is straightforward, and indicates that the optical field $u_j(z+\Delta z)$ can be computed from field values $u_{j-1}(z)$, $u_j(z)$ and $u_{j+1}(z)$ at a given position z. This method based on finite differences is accurate to the first order. Moreover, from a numerical point of view it is a conditionally stable method, where the stability condition is given by:

$$\Delta z \leqslant \Delta x^2/2K = \Delta x^2 n_0\pi/\lambda \quad (5.21)$$

Unfortunately, the value of Δz necessary to assure stability is too small from a practical point of view. As an example, for a mesh of $\Delta x = 0.1$ μm, and a wavelength of $\lambda = 1$ μm, using a reference refractive index of $n_0 = 1.5$, we obtain that the propagation step must be less than $\Delta z < 0.05$ μm. In order to model a 5 mm length device, we would need 10^5 steps!

An alternative way to overcome this problem consists in using a finite difference scheme somewhat similar to the former, known as "*backward-difference*" [5]. The Helmholtz scalar equation expressed in this way takes the following form:

$$2iK\frac{u_j(z+\Delta z)-u_j(z)}{\Delta z}=\frac{u_{j-1}(z+\Delta z)-2u_j(z+\Delta z)+u_{j+1}(z+\Delta z)}{\Delta x^2}+k_0^2(n^2-n_0^2)u_j(z+\Delta z) \quad (5.22)$$

This method has the advantage of being unconditionally stable, although the approximated solution obtained in the simulation is similar to the "forward-difference" method, and thus, no more accuracy is gained.

Fortunately, there is a method, also based on finite difference schemes, that is not only unconditionally stable, but also provides more accurate solutions than the two previous methods. This method, called *Crank-Nicolson scheme* [5], is a linear combination of the "forward difference" method and the "backward-difference" method (Figure 5.5).

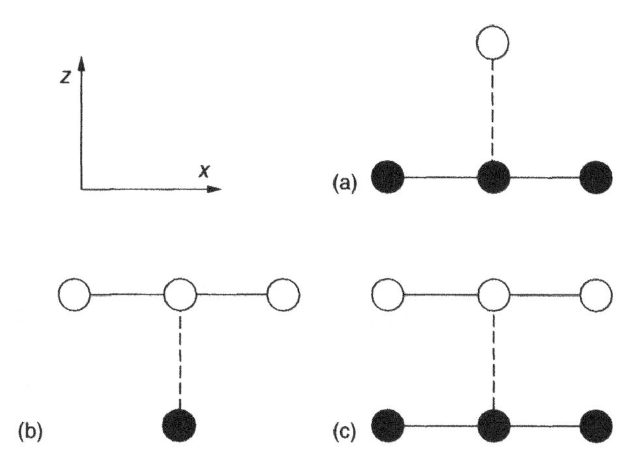

Figure 5.5 Diagram showing the three schemes used in the simulation of light propagation based on finite differences. (a) "Forward" or "fully explicit" is first order accurate, but is stable only for small propagation steps. (b) "Backward" or "fully implicit" is stable for arbitrarily long propagation steps, but is only first-order accurate. (c) "Crank-Nicolson" is second-order accurate, and also is stable for long propagation steps

The finite difference method following a Crank-Nicolson scheme for solving the paraxial propagation equation can be represented as:

$$[2K + i\Delta z\alpha H]u(z + \Delta z) = [2K - i\Delta z(1 - \alpha)H]u(z) \tag{5.23}$$

where the operator H is defined as:

$$Hu \equiv \frac{u_{j-1} - 2u_j + u_{j+1}}{\Delta x^2} + (n_j^2 - n_0^2)k_0^2 u_j \tag{5.24}$$

Expanding this scheme in terms of finite differences, the following equation is obtained:

$$\begin{aligned} 2iK[u_j(z + \Delta z) - u_j(z)] &= k_0^2(n^2 - n_0^2)[\alpha u_j(z + \Delta z) - (1 - \alpha)u_j(z)]\Delta z \\ &+ \left[\alpha \frac{u_{j-1}(z + \Delta z) - 2u_j(z + \Delta z) + u_{j+1}(z + \Delta z)}{\Delta x^2}\right. \\ &\left. - (1 - \alpha)\frac{u_{j-1}(z) - 2u_j(z) + u_{j+1}(z)}{\Delta x^2}\right]\Delta z \end{aligned} \tag{5.25}$$

This equation relates the optical field at $z + \Delta z$, that is $u(z + \Delta z)$, with the field at z, that is, $u(z)$. Rearranging terms in the previous equation, one obtains:

$$a_j u_{j-1}(z + \Delta z) + b_j u_j(z + \Delta z) + c_j u_{j+1}(z + \Delta z) = r_j(z) \tag{5.26}$$

where the a_j, b_j, c_j and r_j coefficients are defined by:

$$a_j = -\alpha \frac{\Delta z}{\Delta x^2}$$

$$b_j = 2\alpha \frac{\Delta z}{\Delta x^2} - \alpha \Delta z[n_j^2(z + \Delta z) - n_0^2]k_0^2 + 2iK$$

$$c_j = -\alpha \frac{\Delta z}{\Delta x^2}$$

$$r_j = (1-\alpha)\frac{\Delta z}{\Delta x^2}\left[u_{j-1}(z) + u_{j+1}(z)\right] +$$
$$\left\{(1-\alpha)\Delta z\left[n_j^2(z) - n_0^2\right]k_0^2 - 2(1-\alpha)\frac{\Delta z}{\Delta x^2} + 2iK\right\}u_j(z) \qquad (5.27)$$

Equation (5.26), besides the coefficients defined by the expression (5.27), forms in fact a *tridiagonal system* of N linear equations[1] ($j = 1, 2, \ldots, N$), which can be solved very efficiently. In addition, it can be demonstrated that the solution to this equation system shows an excellent numerical stability. The algorithm used for solving this tridiagonal system is the *Thomas Method* [6], which requires a computational time that increases with N, while the time required for obtaining a fast Fourier transform using a grid of N points increased as $N\,log_2\,N$.

Strictly speaking, the Crank-Nicolson scheme is unconditionally stable for $\alpha > 0.5$ if the refractive index is independent of x and z. Nevertheless, if the refractive index varies slowly, or if it is uniform with small discontinuities, the Crank-Nicolson method can be applied locally. Under these circumstances, the analysis leads to valid solutions, even for the most adverse situations.

Apart from the numerical stability, the greatest advantage of the Crank-Nicolson method comes from the fact that it provides a better approximation to the exact solution of the problem. While the simple finite difference method (implicit standard scheme) allows a first-order approximation in the propagation step $O(\Delta z + \Delta x^2)$, the Crank-Nicolson method establishes a second-order approximation in the propagation $O(\Delta z^2 + \Delta x^2)$. Therefore, the finite difference method is a powerful numerical method which allows the use of large propagation steps, with the consequent saving in computational time.

An example of light propagation using FD-BPM is shown in Figure 5.6, which corresponds to the same waveguide structure described in Figure 5.4. The field profiles obtained using FD-BPM are quite similar to those obtained when the FFT-BPM

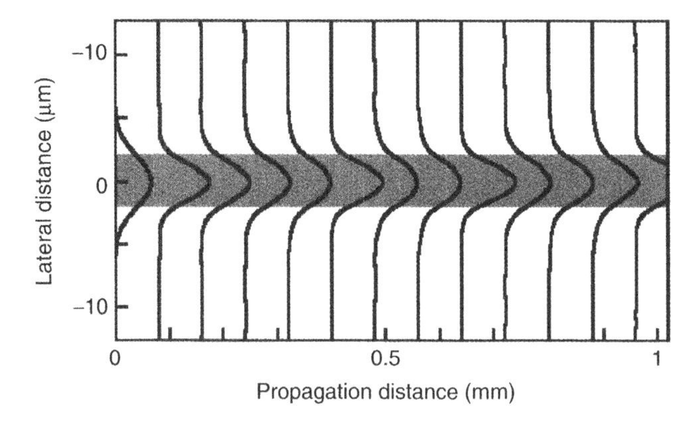

Figure 5.6 FD-BPM propagation based on a Crank-Nicolson scheme with $\alpha = 0.501$. The planar waveguide has the same set of parameters to those described in Figure 5.4, as well as the same parameters used for numerical simulation

technique is applied to simulate light propagation through a straight waveguide. The only difference is that in the FD-BPM simulation the field profiles do not shown any significant noise, although the simulation parameters with respect to that FFT-BPM have been maintained, in particular, the propagation steps' length.

In order to compare the differences between FFT and FD BPM in more detail, we have simulated the propagation of light in slightly more complex structure. In particular we have chosen propagation through a symmetric Y-branch, used in integrated photonic circuits to split the signal into two waveguides, each of them carrying 50% of the input power (Figure 5.7).

In the upper part of Figure 5.7 the results of the simulation using FFT-BPM are shown, and in the lower part the corresponding field profiles obtained using FD-BPM have been also plotted for comparison purposes. Both pictures show the splitting of

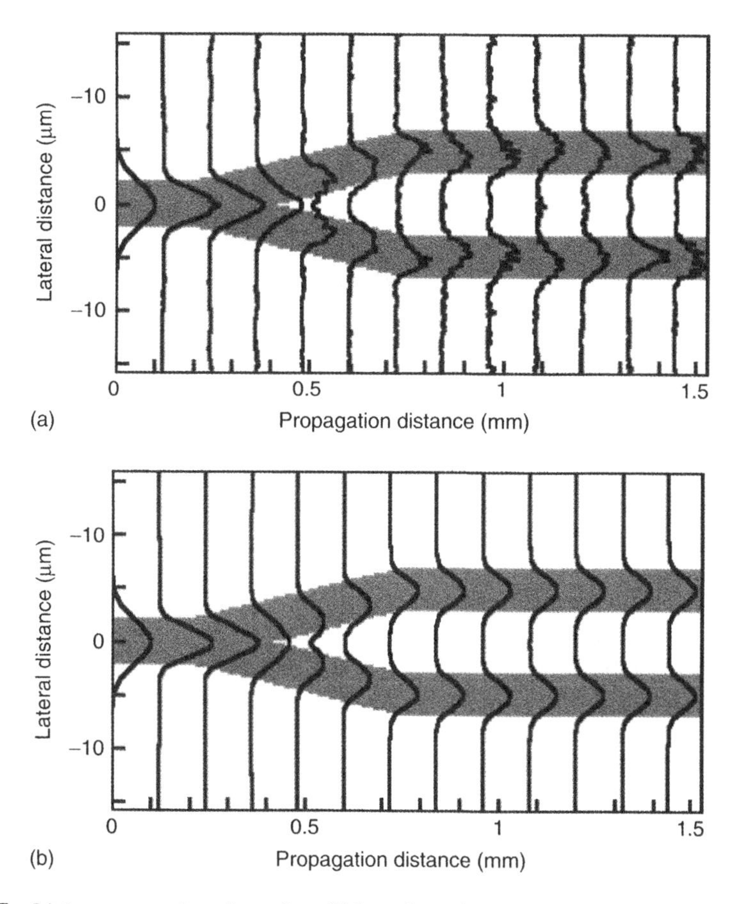

Figure 5.7 Light propagation through a Y-branch, using FFT-BPM (a) and FD-BPM (b). The structure consists initially of a symmetric planar waveguide in step, with a length of 200 μm, which splits into two branches of 550 μm length; from a distance of 750 μm the two branches become parallel. The structure parameters are the following: substrate refractive index 1.5; waveguide refractive index 1.51; waveguide width 4 μm; half-angle branches 0.5°. Simulation parameters: computational window of 256 (2^8) mesh points, separated $\Delta x = 0.25$ μm ($L = 32$ μm); propagation step $\Delta z = 3$ μm; wavelength $\lambda = 1$ μm

light when it enters the Y-branch, and once the branches are separated enough, the field profiles remain almost unchanged. Therefore, the overall behaviour of the device is well described using both numerical methods, although there is a clear difference in the field pattern obtained: while FFT-BPM gives field profiles with ripples, FD-BPM shows very clean field profiles, without any sign of perturbation. Under these circumstances, the choice of FD-BPM to simulate integrated photonic circuits is clear, since it admits larger propagation steps, and also behaves quite well managing structures with large discontinuities in the refractive index. This is not the only advantage of FD over FFT-BPM: if the computational window is reduced in such a way that the optical field reaches the computational boundaries, it is necessary to implement additional algorithms to avoid the optical fields reflecting at the boundaries and re-entering the computational window; otherwise, the simulation of the optical devices will not be correct. This problem can be adequately solved in the case of FD-BPM as we will describe in Section 5.5, but it is cumbersome when using FFT-BPM. An additional advantage of FD-BPM, besides the lower memory and time consumption in modelling complex structures, is the possibility of incorporating wide-angle propagation and full vector algorithms [7, 8].

5.4 Boundary Conditions

Due to the fact that the computational domain in BPM calculations is finite, it is necessary to specify boundary conditions for the optical field at the limits of the computational window. These boundary conditions must be adequately chosen, in such a way that the effect of the boundaries does not introduce errors in the propagation description of the optical field. If these conditions are not well specified, the radiation tends to reflects on the limits of the computational window and comes back to the region of interest, and unwanted interference is produced when the propagation is performed by FD-BPM. In the case of wave propagation based on FFT-BPM, the result is the disappearance of the optical field through a boundary, but the appearance of a new perturbation from the opposite window boundary.

In order to visualise these effects, let us analyse the behaviour of a 3° tilted Gaussian beam propagating in a one-dimensional homogeneous medium. As the Gaussian beam propagates, besides a lateral displacement due to the incident angle, the beam spreads transversally because of the diffraction. If the computational region is wide enough, and therefore the beam does not reach its boundary, the simulation of the beam propagation proceeds normally. On the other hand, if the optical field reaches one of the computational window limits, it will be necessary to impose realistic and reasonably physical boundary conditions for describing and adequately simulating the optical propagation.

Figure 5.8 shows the intensity profiles for the optical field corresponding to the tilted Gaussian beam as it advances along the propagation direction, computed by means of FFT-BPM. When the radiation reaches the upper limit of the window the wave disappears, but at the lower boundary a new perturbation appears, which of course does not correspond to a realistic physical situation. Indeed, one would expect something similar, because propagation by FFT mathematically imposes energy conservation in the computational region.

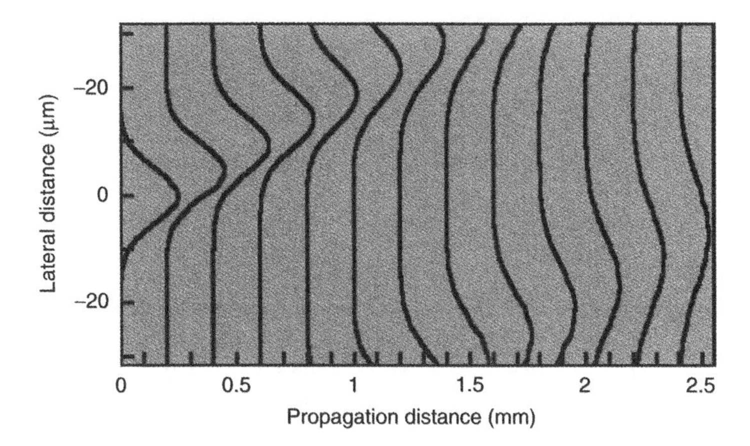

Figure 5.8 FFT-based propagation of a 10 μm width Gaussian beam of $\lambda = 0.5$ μm, which enters with an angle of 3° in a homogeneous medium having a refractive index $n_0 = 2.2$. Parameters of the simulation: longitudinal step $\Delta z = 5$ μm; grid size $\Delta x = 0.4$ μm; 128 mesh points in the transversal direction

An obvious solution to prevent the window limits from interfering in the correct optical propagation is to build a wide enough computational region. In this way, the optical field at the window limits is almost negligible, thus avoiding this unwanted effect. Nevertheless, this procedure implies an unnecessary increment of the window size, and the computing of the optical field in regions out of interest for the solution of the problem.

The most common way of preventing boundary reflection using FFT-BPM is the insertion of artificial absorption regions adjacent to the pertinent boundaries [9]. Usually, the absorption coefficient is ramped from zero at the region's leading edge to some maximum value at the boundary node. Thus, the thickness of the region, the maximum absorption coefficient, and the functional shape must all be carefully chosen for the method to work properly. However, if the gradient in the absorption coefficient is too large, that gradient itself will generate reflections. Although the artificial absorption procedure is accurate provided that the absorption region is adequately tailored, ensuring that this condition is fulfilled for each new problem is often a difficult and time-consuming process. Even when successful, the addition of extra problem zones results in computational penalties of run-time and storage space.

When using FD-BPM for simulating optical propagation we find that the field at points $j = 0$ and $j = N + 1$ are not defined, but are necessary for calculating the field in the interior points ($j = 1$ to N), and we need two extra equations to determine them. *Dirichlet* boundary conditions provide the simplest possibility by specifying the boundary values u_0 and u_{N+1} directly, for instance by setting their values to zero; other possibilities are the *Neumann* or even the *periodic* boundary conditions [5].

Unfortunately, none of these boundary conditions gives satisfactory results, and the implementation causes optical field "reflections" at the window limits, because the condition of zero field at the boundaries is not realistic when the optical perturbation reaches the limits of the computational window. This effect can be appreciated in Figure 5.9 where Dirichlet boundary conditions have been implemented in FD-BPM

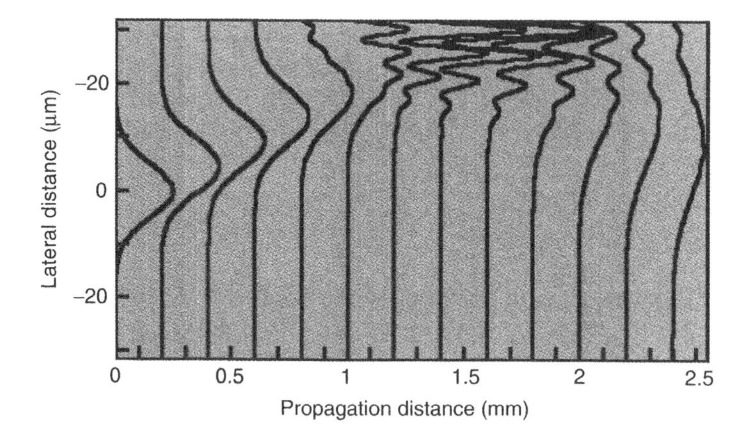

Figure 5.9 FD-BPM simulation for a tilted Gaussian beam propagating through a homogeneous medium (same parameters as in Figure 5.8). When the radiation reaches the upper limit, it suffers a kind of deformation and then bounces and re-enters the domain

simulation: when the Gaussian beam reaches the upper window limit, it initially suffers a deformation, then bounces, and finally returns to the computational region.

One alternative to this dilemma consists in trying to implement realistic boundary conditions from a physical point of view, that is, an algorithm that allows the wave to leave the computational region when it reaches the window limits, without any reflection coming back to the domain. This algorithm is known as *transparent boundary condition* (TBC), and simulates a non-existent boundary [10]. Radiation is allowed to freely escape the problem region without appreciable reflection, whereas radiation flux back into the region is prevented. This TBC employs no adjustable parameters, and thus is problem independent, and can be directly applied to any waveguide structure. In addition, it is easily incorporated into a standard Crank-Nicolson differencing scheme in both two and three dimensions, and it is applicable to longitudinally varying structures of importance for integrated photonic devices [10].

5.4.1 Transparent boundary conditions

The technical description of the TBC implementation begins by considering the scalar paraxial beam-propagation equation (equation (5.6)), and we will focus the discussion on one-dimensional propagation for the sake of simplicity. Since only the boundary region is of interest, we further restrict ourselves to the diffraction terms:

$$\frac{\partial u}{\partial z} = \frac{i}{2K}\frac{\partial^2 u}{\partial x^2} \tag{5.28}$$

where an additional $exp(iKz)$ dependence was assumed in the derivation of equation (5.28). By simple manipulation, equation (5.28) may be rewritten in an energy conservation equation:

$$\frac{\partial}{\partial z}\int_{\mathrm{b}}^{\mathrm{a}} |u|^2 dx = \frac{i}{2K}\left(u^*\frac{\partial u}{\partial x} - u\frac{\partial u^*}{\partial x}\right)\Bigg|_{\mathrm{a}}^{\mathrm{b}} \equiv -F_{\mathrm{b}} + F_{\mathrm{a}} \tag{5.29}$$

where F_b represents the energy flux leaving the right boundary and F_a represents that entering through the left boundary. Since the treatment of the two boundaries is essentially identical, we consider only the right boundary. We next make the important assumption that at this boundary the field is of the form:

$$u = u_0 e^{ik_x x} \tag{5.30}$$

where u_0 and k_x are in general complex, and k_x is for the moment unknown. With this assumption, the flux F_b becomes:

$$F_b = \frac{Re(k_x)|u(b)|^2}{K} \tag{5.31}$$

where Re indicates the real part. Therefore, as long as the real part of k_x is positive, the contribution to the overall change in energy from this boundary will always be negative, i.e., radiative energy can only flow out of the problem region.

Within the Crank-Nicolson scheme based on finite differences, and assuming the same exponential dependence described above, the optical field in the limit of the window u_N^j prior to the start of the $(j+1)$th propagation step should fulfil the following relation:

$$\frac{u_N^j}{u_{N-1}^j} = \frac{u_{N-1}^j}{u_{N-2}^j} = e^{ik_x \Delta x} \tag{5.32}$$

This expression allows the determination of k_x, after completion of the jth step, using two interior points close to the boundary. Then, the boundary condition for the next propagation step $(j+1)$ is thus:

$$u_N^{j+1} = u_{N-1}^{j+1} \cdot e^{ik_x \Delta x} \tag{5.33}$$

where the k_x value is the previously calculated using equation (5.32) in the jth step. However, prior to the application of equation (5.33), the real part of k_x must be restricted to be positive to ensure only radiation outflow. Therefore, if from equation (5.32) the real part of k_x is negative, it is reset to zero, and the field at the boundary is redefined using the value of k_x just calculated.

Since the boundary condition itself is linear and only involves the two mesh points nearest the boundary, the ratio between the interior points previously used to determine k_x is allowed to change. Thus, the value of k_x computed for the next propagation step will generally be different. Such an adaptive procedure is required if an accurate algorithm that reflects the minimum amount of energy back into the problem region is to be constructed.

The great advantage of the TBC method lies in its convenience and usefulness. While the use of artificial absorbers to remove scattered radiation is clumsy, imposes some penalty of computer run-time and storage, and must be retailored for each new problem, the TBC algorithm uses no adjustable parameters, thus being problem independent, and imposes no storage penalty, because it requires no extra computational zones to be considered.

Besides its accuracy and efficiency, the TBC algorithm shows a high degree of robustness. The implementation is easily performed for one-dimensional problems as well as for propagation through 2D structures. In each case, paraxial propagation leads

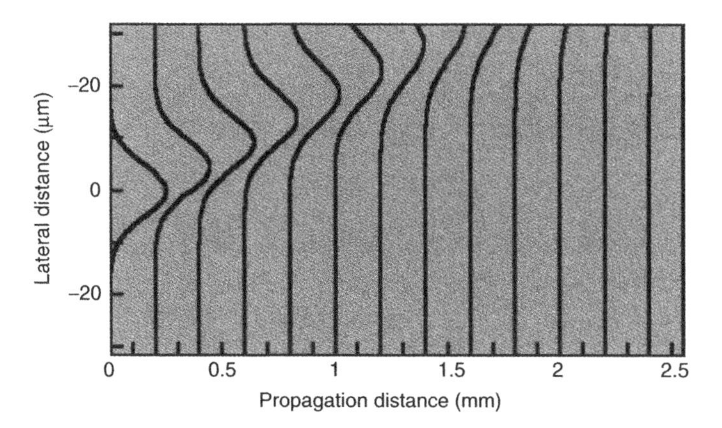

Figure 5.10 Tilted Gaussian beam propagation by means of FD-BPM, using the same parameters as in Figures 5.8 and 5.9, when TBC is implemented. When the radiation reaches the upper window limit, it passes through the boundary and virtually disappears, leaving the computational region

to a solution of a tridiagonal system, and thus the inclusion of TBC does not make the resolution of the problem more complicated.

In order to show the accuracy and robustness of TBC, we have simulated the propagation of a Gaussian beam with the same set of parameters used in Figures 5.8 and 5.9, where we observed incorrect simulation when the radiation reached the upper limit of the window. Figure 5.10 shows the propagation of the 3° tilted Gaussian beam through an homogeneous medium simulated by the FD-BPM technique, where we have implemented the TBC algorithm in both the upper and lower boundaries. We observe that when the radiation reaches the upper window limit, it virtually disappears into the boundary, without any sign of reflection or distortion in the beam shape as it passes through the boundary. Indeed, the total disappearance of the beam as it crosses the limit of the computational region is the expected result from a physical point of view. The effective reflection coefficient, defined as the ratio between the initial energy in the computational region and the energy in it after a long enough propagation distance, is in this case lower than 10^{-4}, demonstrating the high efficiency and accuracy of the TBC method for simulating optical propagation based on FD-BPM.

5.5 Spatial Frequencies Filtering

In many circumstances, the simulation of optical propagation in waveguides gives rise to the appearance of rippling in the electric field pattern after a certain propagation distance, due to numerical noise. This effect is produced in both FFT and FD-BPM algorithms. In order to correct this perturbation, which does not originate in any real physical mechanism, it is necessary to resort to frequency filtering techniques in the spatial frequency domain, using low-pass digital filters. In order to do this, the spatial frequency spectrum of the electric field distribution is first calculated at a given point in the propagation, and then it is multiplied by a certain *window function*, in such a way that the high frequency components, which have a low signal to noise ratio and the beam carries a small percentage of the total power, are removed (or at least reduced), and the low frequencies components where the higher amount of optical field power is concentrated are retained. It is necessary to remember that filtering windows must be used with caution, because some essential frequency components of the optical field which could be necessary for the correct description of the optical propagation can be eliminated, and thus the simulation would lose accuracy. In general, the filtering window is chosen in such a way that it includes at least 95% of the energy transported by the optical field.

Figure 5.11 represents the evolution of the spatial frequency power spectrum associated to light propagation through a step-index planar waveguide, using the FFT-BPM algorithm with a propagation step of $\Delta z = 5$ μm. We can observe the appearance of high frequency components that give rise to ripples in the optical field, as can be seen on the right-hand side of Figure 5.11. This effect is much more pronounced as the propagation step increases, and it is more important when the waveguide profile shows high index differences.

In order to avoid or reduce noise in optical field simulations, it is necessary to use smaller propagation steps, which involves increasing the computational time, or alternatively implementing filtering techniques.

In this particular example, we have chosen a filter window that allows frequencies lower than 0.025 μm^{-1} to pass through, and eliminates the portion of the frequency spectrum with frequencies higher that 0.025 μm^{-1}, allowing 95% of the optical power to be included in the non-filtered components. The resulting frequency spectrum is shown in Figure 5.12 (a), along with the optical power profile obtained after using

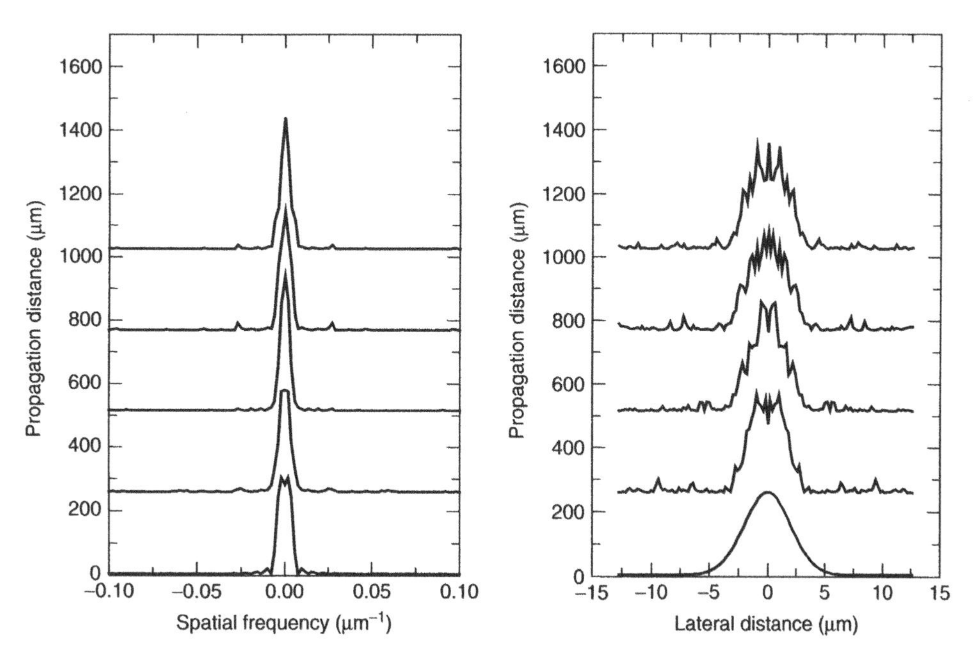

Figure 5.11 Left: evolution of the spatial frequency power spectrum associated with propagation in a step-index planar waveguide of 8 μm width, using a propagation step of $\Delta z = 5$ μm. Right: optical power profiles corresponding to propagation in the planar waveguide, where the appearance of ripples in the optical field is clearly observed

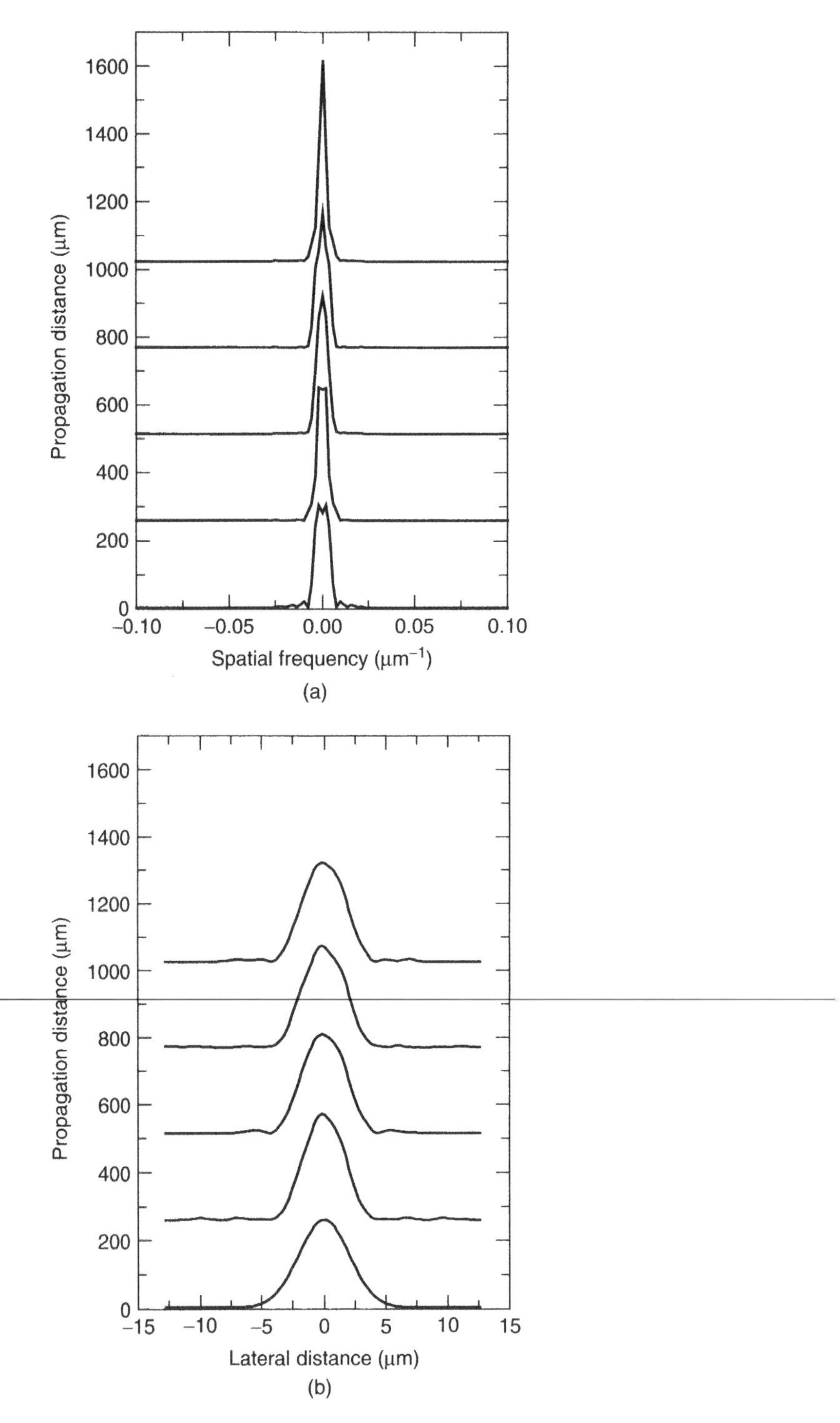

Figure 5.12 (a): frequency spectrum after filtering with a window that removes spatial frequencies higher than 0.025 μm^{-1}. (b): optical power profile obtained after propagating the field by applying filtering techniques, where we observe the disappearance of the ripples in the optical field

this low-pass filter. From Figure 5.12 it can be seen that the ripples that appeared in the optical field profile when the propagation was done without filtering have now disappeared, and the optical field obtained after using filtering techniques shows a profile clean of noise or ripples.

The incorporation of filtering techniques in optical propagation algorithms is very easy to implement, in both FFT and FD-BPM, and there is almost no penalty in computing time. As we have shown above, the implementation of these techniques allows the use of longer propagation steps in the optical simulation, thus saving a lot of CPU time. This situation is particularly of interest in simulating 2D structures such as channel waveguides, where the appearance of ripples in the optical field is immediate unless very short propagation steps are used.

5.6 Modal Description Based on BPM

BPM offers an accurate description of spatial (near-field) and angular (far-field) properties of the electric field, but it can also generate information relevant to a purely modal description of the field [11].

In particular, the mode propagation constants and the power associated with each mode can be determined from a Fourier analysis of the *complex field amplitude correlation function* defined by $P(z) = \langle u^*(0)u(z)\rangle$, where the brackets signify integration over the waveguide cross-section, and z represents axial distance. The function $P(\beta)$, which is the Fourier transform with respect to z of $P(z)$, displays a set of resonant peaks that identify the guided modes (and also the radiation modes) that have been excited by the input source at $z = 0$. The peaks occur at values of β that correspond to the mode propagation constants β_j, and the heights of the peaks are proportional to the powers of the corresponding modes.

In order to see how these results are obtained, let us consider light propagation in a straight waveguide, in which the refractive index has only dependence on the x and y coordinates, $n = n(x, y)$. The solution for the scalar Helmholtz equation can be approximated by the product of a complex field amplitude $u(x, y, z)$ and a carrier wave moving in the positive z-direction: $E(x, y, z) = u(x, y, z)e^{-iKz}$, where $K = n_0\omega/c$, and n_0 is the reference refractive index.

In addition, the complex field amplitude $u(x, y, z)$ can be expressed in terms of the waveguide-mode eigenfunctions as:

$$u(x, y, z) = \sum_{\nu} A_\nu f_\nu(x, y)e^{-i\beta_\nu z} \tag{5.34}$$

where β_ν is the propagation constant of the νth mode, $f_\nu(x, y)$ is its electric field distribution in the transversal direction, and A_ν is its modal weight associated to the power it carries. It is important to remark here that, using this notation, the real propagation constant $\beta_\nu^{'}$ of the modes is related to the values of β_ν (from equation 5.34) by:

$$\beta_\nu^{'} = \beta_\nu + K \tag{5.35}$$

The procedure for determining the mode weights and mode constants (modal spectrum) of a straight waveguide is called the *correlation-function method* [11]. For this purpose, let us form the product $u^*(x, y, 0)u(x, y, z)$ and integrate over the cross-section of the waveguide. Making use of equation (5.34), we obtain for the correlation function $P(z)$:

$$P(z) = \int_{-\infty}^{\infty}\int_{-\infty}^{\infty} u^*(x, y, 0) \bullet u(x, y, z) dx dy = \langle u^*(0)u(z)\rangle \tag{5.36}$$

Bearing in mind the mode orthogonality, and assuming that the transversal distributions of the modes are normalised:

$$\int_{-\infty}^{\infty}\int_{-\infty}^{\infty} f_\mu^* f_\nu dx dy = \delta_{\mu\nu} \tag{5.37}$$

the correlation function $P(z)$ can be alternatively expressed as a function of the modal weights A_ν and the propagation constants β_ν in the form:

$$P(z) = \sum_\nu |A_\nu|^2 e^{-i\beta_\nu z} \tag{5.38}$$

Taking the Fourier transform of equation (5.38) gives:

$$P(\beta) = \sum_\nu |A_\nu|^2 \delta(\beta - \beta_\nu) \tag{5.39}$$

where δ is the delta Kronecker function. This expression suggests that the calculated spectrum of $P(z)$ (that is, its Fourier transform) will display a series of resonances with maxima at $\beta = \beta_\nu$, and peak values proportional to the mode weight coefficients A_ν. The coefficient A_ν is related to the relative power W_ν carried by the νth mode by:

$$W_\nu = |A_\nu|^2 \tag{5.40}$$

being:

$$W = \sum_\nu W_\nu \tag{5.41}$$

the total power carried in the waveguide.

In practice, as only a finite record of $P(z)$ is available, the resulting resonances in the spectrum $P(\beta)$ will thus exhibit a finite width and shape that are characteristic of the record length Z. Since in general the resonance peaks do not coincide exactly with the sampled values of β, errors will result in the values of W_ν and β_ν values inferred from the maxima in sampled data set for $P(\beta)$. The maximum uncertainty in β_ν can be expressed in terms of the sampling interval $\Delta\beta$ along the β axis and the propagation distance Z over which the solution $u(x, y, z)$ is available as:

$$\Delta\beta_\nu = \Delta\beta/2 = \pi/Z \tag{5.42}$$

In order to reduce the uncertainty in the determination of the propagation constant β_ν it is necessary to increase the propagation length Z, but this at the expense of an increase in the computational time. It is possible, however, to improve substantially on the accuracy implied in equation (5.42) by multiplying the data sample to be Fourier transformed by a suitable window function and then selecting the β_ν values from the transformed sample by interpolation, or even by a line-shape fit for the individual resonances.

On the other hand, the magnitudes of the propagation constants for the guided modes will be limited according to:

$$0 < |\beta_\nu| < K\Delta n_{max} \tag{5.43}$$

where $\Delta n_{max} = n_{max} - n_0$. Therefore, to ensure that the axial spectrum is accurately represented it is necessary for the axial sampling distance Δz to satisfy:

$$\Delta z < \pi/K\Delta n_{max} \tag{5.44}$$

or expressed as function of the wavelength:

$$\Delta z < \lambda/2n_0\Delta n_{max} \tag{5.45}$$

In practice, this condition is satisfied by a factor of ~ 5.

Let us now see how to proceed in practice to obtain the propagation constants of the modes, as well as their modal weights. For this purpose, let us consider a one-dimensional multi-mode waveguide which is invariant in the propagation direction (straight waveguide). If the light excitation at the input coincides with the field distribution corresponding to a guided mode, we will expect this light distribution to propagate without losing its shape and intensity, because the energy transfer between modes is forbidden by the modal orthogonality relationship. If, however, the light injection at $z = 0$ provokes the excitation of various modes, the transversal distribution of the optical field will change as the beam propagates along the waveguide due to the fact that each mode has a different propagation constant, and therefore the relative phase between modes will change as a function of the propagation distance.

Figure 5.13 shows the evolution of the transversal light distribution in a one-dimensional straight waveguide when a Gaussian beam with constant phase is launched

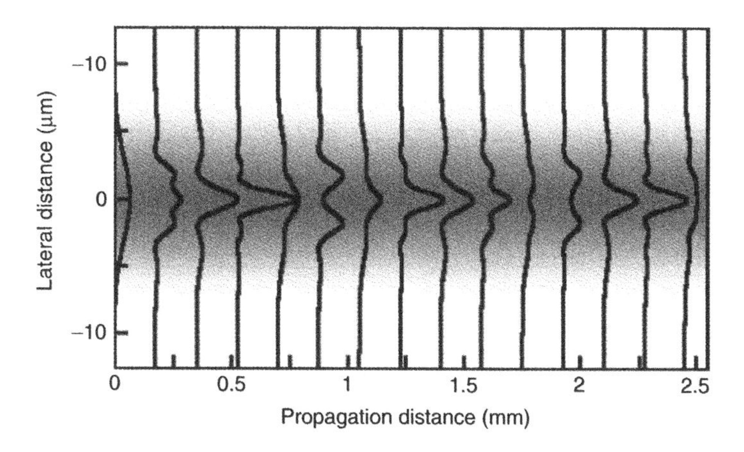

Figure 5.13 Light propagation, using FD-BPM with TBC, in a multi-mode straight waveguide. At the input a Gaussian beam of 6 μm width centred at $x = 0$ and constant phase is injected. The waveguide consists of a symmetric planar waveguide having a Gaussian refractive index profile of 8 μm, maximum index difference $\Delta n = 0.03$ and a substrate refractive index of $n_0 = 2.2$. Parameters of the simulation: 128 point in the transversal direction; grid size $\Delta x = 0.2$ μm; propagation step $\Delta z = 5$ μm; wavelength $\lambda = 1$ μm

at the input position $z = 0$. The planar waveguide has a symmetric Gaussian refractive index profile with 6 μm width and $\Delta n = 0.03$ with respect to the substrate index of $n_0 = 2.20$. The wavelength used is $\lambda = 1$ μm, and the simulation has been performed by FD-BPM, where TBC has been implemented. For this wavelength, the waveguide structure supports six guided modes. As can be seen in Figure 13, the field profile changes as the beam propagates, due to the fact that the light distribution at the input does not correspond to any particular guided mode, but several of them have been simultaneously excited.

Following equation (5.36), we build the correlation function $P(z)$ corresponding to the propagation described in Figure 5.13, and presented in Figure 5.14. After first inspecting this figure it seems difficult to extract any relevant information about the behaviour of the modal propagation; perhaps the only clear fact is that the evolution of $P(z)$ does not follow a pure senoidal function.

Nevertheless, a Fourier transform of the correlation function $P(z)$ gives a much clearer picture, as it can be seen in Figure 5.15. From Figure 5.15 it is possible to observe a series of well-defined peaks, ranging between 0.0 and 0.16 μm^{-1}, with different heights.

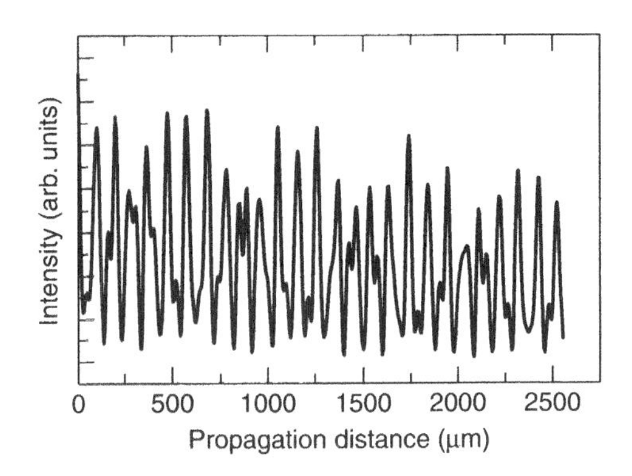

Figure 5.14 Correlation function $P(z)$ corresponding to the propagation described in Figure 5.13

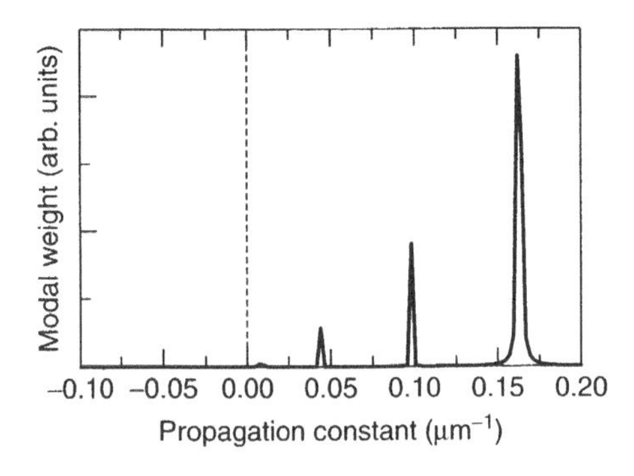

Figure 5.15 Axial spectrum obtained from a 2560 μm propagation length in a planar waveguide, after launching Gaussian profile input centred at $x = 0$. Only the even (symmetric) modes transport energy

Table 5.1 Propagation constants determined by the correlation-function method after two distinct excitations, besides their "exact" values calculated by the multi-layer method

Mode	$\beta_\nu(1)$ (μm^{-1})	$\beta_\nu(2)$ (μm^{-1})	β_ν (Exact) (μm^{-1})	Error (μm^{-1})
0	0.1620	0.1620	0.1684	0.0064
1	–	0.1301	0.1311	0.0001
2	0.0982	0.0982	0.0971	0.0011
3	–	0.0687	0.0671	0.0016
4	0.0442	0.0442	0.0413	0.0029
5	–	0.0245	0.0203	0.0042
6	0.0074	0.0074	0.0044	0.0030

In Figure 5.15 the resonance peaks appear at values of the propagation constant $\beta = 0.1620, 0.0982, 0.0442$ and $0.0074\ \mu\text{m}^{-1}$, with different intensities increasing with β. These values correspond to guided modes of the structure, because $\beta > 0$. Table 5.1 shows these values $\beta_\nu(1)$, besides the "exact" propagation constants obtained using a multi-layer analysis, as was described in Chapter 3. From Table 5.1 it is evident that only the even modes have been excited by the symmetric Gaussian injection, centred at $x = 0$, at the input of the waveguide. On the other hand, the error found in the propagation constants, referred to their exact values, indicates that it is a bit higher that the expected one calculated by the formula (5.43):

$$\Delta\beta_j = \Delta\beta/2 = \pi/Z = 0.002\ \mu\text{m}^{-1}$$

This small discrepancy is due to the discretisation of the structure, in this particular case a Gaussian refractive index profile.

In addition, the use of a propagation step of $\Delta z = 5\ \mu$m assures that the whole modal spectrum is represented, because the condition:

$$\Delta z < \lambda/2n_0\Delta n_{\text{max}} = 7.6\ \mu\text{m}$$

is fulfilled.

In order to obtain the propagation constant of every single mode, it is necessary that the light injection at the input excites adequately each of them. Choosing the same Gaussian excitation, but now shifted 6 μm respect to the waveguide centre, we obtain the propagation depicted in Figure 5.16.

The modal spectrum corresponding to this propagation is represented in Figure 5.17, where now it can be observed that the six propagating modes supported by the waveguide have been excited, although with different amount of energy. The propagation constants $\beta_\nu(2)$ calculated from this graph have been included in Table 5.1, where we observe that they coincide with the analysis performed with a different excitation. It should be also pointed out that a small fraction of energy is presented for negative values of the propagation constant. This fact indicates that the electric field at the input has also excited radiation modes.

5.6.1 Modal field calculation using BPM

Once the propagation constants of the waveguide modes have been calculated, with a further and simple analysis it is also possible to determine the transversal field

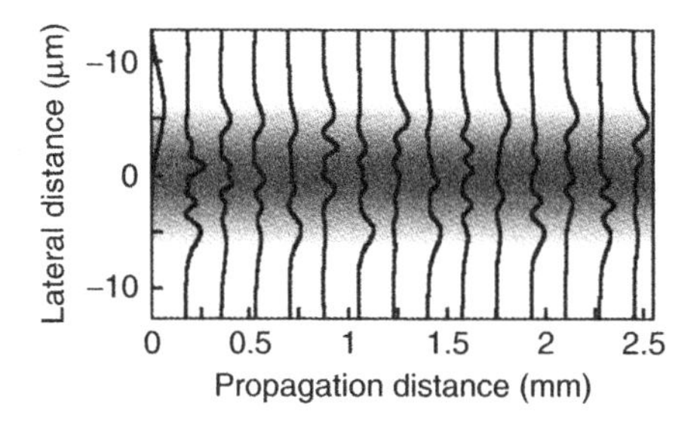

Figure 5.16 FD-BPM simulation in a symmetric multi-mode waveguide, after the light injection having a Gaussian profile of 6 μm width centred at $x = 6$ μm, and constant phase. The waveguide structure, as well as the simulation parameters, are similar to those presented in Figure 5.13

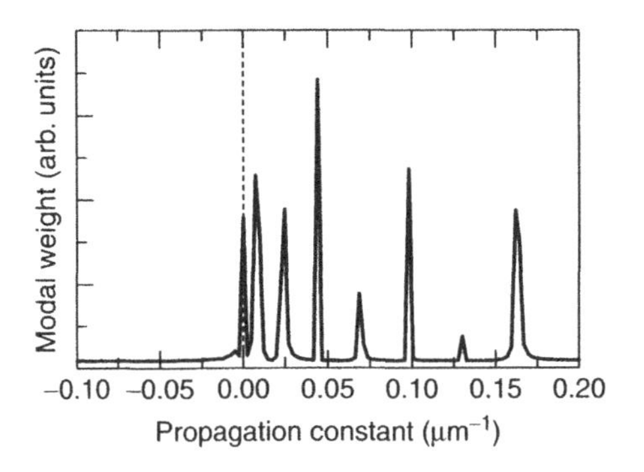

Figure 5.17 Axial spectrum obtained after launching a shifted Gaussian beam. This input source is able to excite all modes, as well as radiation modes

distribution $f(x, y)$ corresponding to each propagation mode. For this purpose, it is enough to allow the input source to propagate along the straight waveguide, and then multiply the optical field simulated at each propagation step by the factor $exp(i\beta_\mu z)$, where β_μ indicates the propagation constant (previously calculated) of the μth mode whose transverse field distribution $f_\mu(x, y)$ we want to obtain. In such a way, if we now add these contributions as the wave propagates, all the modes, except the μth mode, will have a phase which changes with the propagation distance, and will cancel for a long enough propagation length. In contrast, the μth mode will not have any z-dependence, because it has been eliminated after multiplying the optical field by the factor $exp(i\beta_\mu z)$. Therefore, the transversal field distribution corresponding to the μth mode will build up as the perturbation advances along the straight waveguide.

We restrict the problem to a one-dimensional waveguide, and therefore the field profiles will have dependence only on the x coordinate. If we assume that the optical field in the waveguide can be expressed only by the set of m confined modes, thus ignoring the energy carried out by the continuum radiation modes, the optical field can be written as:

$$u(x, z) = \sum_{\nu=0}^{m-1} A_\nu f_\nu(x) e^{-i\beta_\nu z} \tag{5.46}$$

For obtaining the field distribution corresponding to the μth mode, we build the function $G_\mu(x)$ defined as:

$$G_\mu(x) = \int_0^Z u(x,z)e^{i\beta_\mu z}dz = \int_0^Z \left(\sum_{\nu=0}^{m-1} A_\nu f_\nu(x) e^{-i\Delta\beta_{\mu\nu}z} \right) dz \tag{5.47}$$

where Z is the total propagation length, and $\Delta\beta_{\mu\nu}$ is defined as $(\beta_\nu - \beta_\mu)$. This function can be separated in to two terms as:

$$G_\mu(x) = A_\mu f_\mu(x) Z + \sum_{\substack{\nu=0 \\ \nu\neq\mu}}^{m-1} \left(A_\nu f_\nu(x) \int_0^Z e^{-i\Delta\beta_{\mu\nu}z}\, dz \right) \tag{5.48}$$

The integrals in equation (5.48) are periodic functions of z, and thus they are limited to a given value, depending on the propagation constant of the modes. In particular, their absolute values are limited to $|2/\Delta\beta_{\mu\nu}|$, and therefore the second term in equation (5.48) is a bounded value, which depends on the power transported for each mode through their modal weights A_ν. By contrary, the first term in equation (5.48) is a function which grows linearly with the propagation length Z. Therefore, for large values on the propagation length, the second term can be ignored, and the first term will reproduce the transverse field distribution of the μth mode, or in other words, the function $G_\mu(x)$ is proportional to the field distribution $f_\mu(x)$. To obtain an accurate field distribution, and assuming equally spaced modes and similar modal weights, the propagation length should fulfil the condition:

$$Z \gg m(m-1)/\Delta\beta \tag{5.49}$$

m being the number of confined modes which supports the waveguide, and $\Delta\beta$ the difference between the propagation constants of two consecutive modes.

Nevertheless, although Z was chosen long, it does not guarantee the total elimination of the contribution of a specific mode ν, and to accurately build the field distribution of a mode it is necessary to excite it with a high fraction of the total input energy. In order to overcome these problems, we can alternatively choose a slight different method.

First, we build the function $H_0(x)$ defined as:

$$H_0(x) = \int_0^{2\pi/\Delta\beta_{\mu 0}} u(x,z)e^{i\beta_\mu z}\, dz \tag{5.50}$$

or expressed in terms of the modal fields:

$$\begin{aligned} H_0(x) &= \int_0^{2\pi/\Delta\beta_{\mu 0}} \left(\sum_{\nu=0}^{m-1} A_\nu f_\nu(x) e^{-i\Delta\beta_{\mu\nu}z} \right) dz = \\ &= \int_0^{2\pi/\Delta\beta_{\mu 0}} A_0 f_0(x) e^{-i\Delta\beta_{\mu 0}z}\, dz \\ &+ \int_0^{2\pi/\Delta\beta_{\mu 0}} \left(\sum_{\nu=1}^{m-1} A_\nu f_\nu(x) e^{-i\Delta\beta_{\mu\nu}z}\, dz \right) \end{aligned} \tag{5.51}$$

The first term in the right hand of the above equation is exactly cancelled, because of the proper choice of the integral limits. Thus, we ensure that the function $H_0(x)$ has no more contribution to the 0th mode field distribution given by $f_0(x)$.

Second, we now choose the function $H_0(x)$ previously calculated as a new input field $u(x, z = 0)$, and we perform a new propagation sequence. The new input field cannot therefore excite the 0th mode. We repeat the steps one and two for each mode, except for the μth mode, obtaining the function $H_1(x)$, $H_2(x)$, etc. In this way, we are subtracting the contribution of each propagating mode to the optical field, and the result is the procurement of the modal profile for the μth mode.

This new method is exact, because all the modes, except the selected one, are eliminated. Also, this procedure is accurate providing that the propagation constants of the propagating modes are well determined, and assuming that the contribution to the radiation modes can be neglected. The main advantage of this method is that the propagation steps can be exactly calculated for the elimination of each mode, and consequently the algorithm is straightforward, thus saving computational time, which is particularly important for modelling two-dimensional structures.

As an example of an application, let us consider a one-dimensional straight waveguide, which is multi-mode at $\lambda = 1.55$ μm. The planar waveguide has a symmetric Gaussian refractive index profile as shown in Figure 5.18, with a width of $d = 3.5$ μm, and a maximum index change of $\Delta n = 0.030$, where the substrate refractive index is $n_0 = 2.200$. Here we have made no distinction between TE and TM modes because of the small and gradual index difference of the waveguide refractive index profile.

Figure 5.19 shows the intensity (square of the optical field) profiles corresponding to the four confined modes which supports the waveguide at $\lambda = 1.55$ μm. The intensity profiles shown in Figure 5.19a correspond to the modal distributions obtained by the FD-BPM algorithm using $\Delta z = 5$ μm length steps. The input source is a Gaussian beam of 5 μm width, shifted 3 μm with respect to the centre of the planar waveguide. The window size is 32 μm, with 128 mesh points size 0.25 μm. Together with the intensity profiles, we have included the propagation constants calculated by the correlation method algorithm, which were used to compute the intensity profiles as well.

In order to compare these results, Figure 5.19b shows the intensity modal profiles of the planar waveguide, but obtained by the multi-layer method described in the

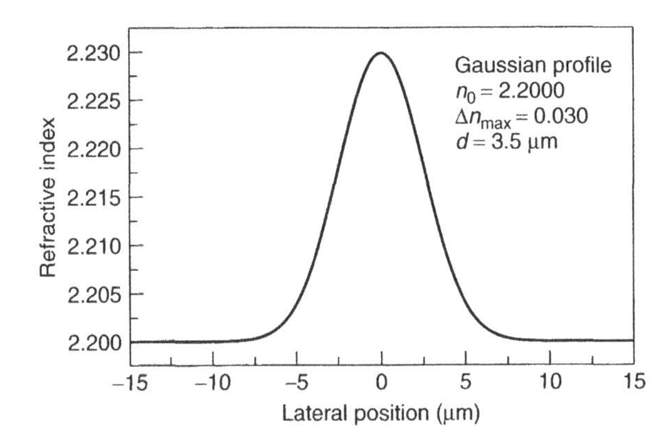

Figure 5.18 Gaussian refractive index profile used to calculate the modal field distributions using BPM algorithm

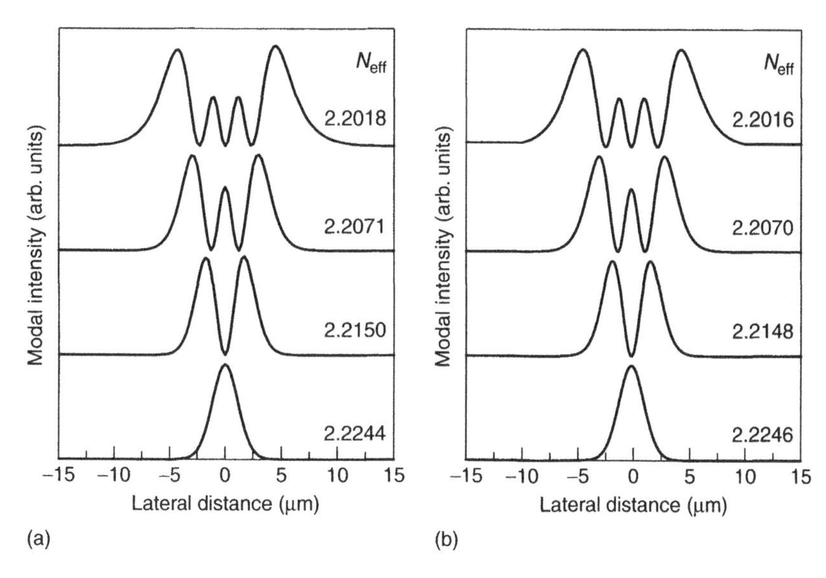

Figure 5.19 Modal intensities corresponding to the four propagating modes in a planar waveguide with graded Gaussian index profile. (a): intensity profiles and propagation constants calculated by the BPM algorithm. (b): results from the multi-layer method

Chapter 3, and the propagation constants of the modes. We can observe that not are only similar values of the propagation constants obtained, but also the intensity profiles coincide accurately with the ones calculated by the BPM algorithm.

Note

[1]All the elements of a tridiagonal matrix are equal to zero, except for the diagonal and the two parallel lines adjacent to it, i.e., $u_{ij} = 0$, except $u_{ii}, u_{i-1,i}$ and $u_{i+1,i}$.

References

[1] M.D. Feit and J.A. Fleck, "Light Propagation in Graded-Index Optical Fibers", *Applied Optics*, **17**, 3990–3998 (1978).

[2] R. Scarmozzino and R.M. Osgood, "Comparison of Finite-Difference and Fourier-Transform Solutions of the Parabolic Wave Equation with Emphasis on Integrated-Optics Applications", *Journal of the Optical Society of America A*, **8**, 724–731 (1991).

[3] W. Huang, C. Xu, S. Chu and K. Chaudhuri, "The Finite-Difference Vector Beam Propagation Method: Analysis and Assessment", *Journal of Lightwave and Technology*, **10**, 295–305 (1992).

[4] W.H. Press, S.A. Teukolsky, W.T. Vetterling and B.P. Flannery, *Numerical Recipes in Fortran 77: The Art of Scientific Computing*, Chapter 12, Cambridge University Press, New York (1996).

[5] W.H. Press, S.A. Teukolsky, W.T. Vetterling and B.P. Flannery, *Numerical Recipes in Fortran 77: The Art of Scientific Computing*, Chapter 19, Cambridge University Press, New York (1996).

[6] W.H. Press, S.A. Teukolsky, W.T. Vetterling and B.P. Flannery, *Numerical Recipes in Fortran 77: The Art of Scientific Computing*, Chapter 2, Cambridge University Press, New York (1996).
[7] G.R. Hadley, "Multistep Method for Wide-Angle Beam Propagation", *Optics Letters*, **17**, 1743–1745 (1992).
[8] M.S. Al Salamed and M.R. Owais, "Full Wave Analysis of Dielectric Optical Waveguides by Vectorial Finite Elements and Absorbing Boundary Condition", *Journal of Optical Communications*, **20**, 74–78 (1999).
[9] D. Yevick and B. Hermansson, "New Formulation of the Matrix Beam Propagation Method: Application to Rib Waveguides", *IEEE Journal of Quantum Electronics*, **25**, 221–229 (1989).
[10] G.R. Hadley, "Transparent Boundary Condition for the Beam Propagation Method", *IEEE Journal of Quantum Electronics*, **28**, 363–370 (1992).
[11] M.D. Feit and J.A. Fleck, "Computation of Mode Properties in Optical Fibers Waveguides by a Propagating Beam Method", *Applied Optics*, **19**, 1154–1164 (1980).

Further Reading

O. Zhuromskyy, M. Lohmeyer, N. Bahlmann, P. Hertel, H. Dötsch and A.F. Popkov, "Analysis of Nonreciprocal Light Propagation in Multimode Imaging Devices", *Optical and Quantum Electronics*, **32**, 885–897 (2000).
S.F. Helfert and R. Pregla, "Determining Reflections for Beam Propagation Algorithms", *Optical and Quantum Electronics*, **33**, 343–358 (2001).
T.P. Felici and D.F.G. Gallagher, "On Propagation through Long Step Tapers", *Optical and Quantum Electronics*, **33**, 399–411 (2001).

Appendix 1

COMPLEX NOTATION OF THE ELECTRIC AND MAGNETIC FIELDS

The electromagnetic field is composed by two vectorial fields, the electric and magnetic fields. In the case of monochromatic waves, the fields have harmonic dependence on time, and thus can be expressed as:

$$\mathcal{E}(\mathbf{r}, t) = \mathcal{E}_0(\mathbf{r}) \cos[\omega t + \varphi(\mathbf{r})]$$

$$\mathcal{H}(\mathbf{r}, t) = \mathcal{H}_0(\mathbf{r}) \cos[\omega t + \varphi(\mathbf{r})]$$

Using complex notation, the electric and magnetic fields are expressed as:

$$\mathcal{E}(\mathbf{r}, t) = Re[\mathbf{E}(\mathbf{r})e^{i\omega t}]$$

$$\mathcal{H}(\mathbf{r}, t) = Re[\mathbf{H}(\mathbf{r})e^{i\omega t}]$$

where $\mathbf{E}(\mathbf{r})$ and $\mathbf{H}(\mathbf{r})$ denote the complex amplitudes of the electric and magnetic fields, respectively.

Taking into account Moivre's formula, $e^{ix} = \cos(x) + i\sin(x)$, and expanding the complex amplitudes $\mathbf{E}(\mathbf{r})$ and $\mathbf{H}(\mathbf{r})$ into their real and imaginary parts:

$$\mathbf{E}(\mathbf{r}) = \mathbf{E}_r(\mathbf{r}) + i\mathbf{E}_i(\mathbf{r})$$

$$\mathbf{H}(\mathbf{r}) = \mathbf{H}_r(\mathbf{r}) + i\mathbf{H}_i(\mathbf{r})$$

the relations between the real and imaginary part of the complex amplitudes with the amplitudes $\mathcal{E}_0(\mathbf{r})$ and $\mathcal{H}_0(\mathbf{r})$ and the initial phase $\varphi(\mathbf{r})$ are given by:

$$\mathbf{E}_r(\mathbf{r}) = \mathcal{E}_0 \cos\varphi(\mathbf{r})$$

$$\mathbf{E}_i(\mathbf{r}) = -\mathcal{E}_0 \sin\varphi(\mathbf{r})$$

$$\mathbf{H}_r(\mathbf{r}) = \mathcal{H}_0 \cos\varphi(\mathbf{r})$$

$$\mathbf{H}_i(\mathbf{r}) = -\mathcal{H}_0 \sin\varphi(\mathbf{r})$$

Appendix 2

PHASE SHIFTS FOR TE AND TM INCIDENCE

Program core for the calculation of the phase shifts suffered for soft and hard incidence, for TE and TM polarised waves.

```
  n1 = Val(Text1.Text)
  n2 = Val(Text2.Text)
  pi = 4 * Atn(1)
  xmin = Val(Text4.Text)
  xmax = Val(Text5.Text)
  steps = Val(Text11.Text)
  ymin = Val(Text7.Text)
  ymax = Val(Text8.Text)
  Picture1.Scale (xmin, -ymax)-(xmax, -ymin)
  Picture1.Line (xmin, -ymax)-(xmax, -ymin), 15, B
  Picture1.Line (xmin, 0)-(xmax, 0), 15
  'Tics in axis
  For t = xmin To xmax Step 10
   Picture1.Line (t, 0)-(t, -5)
  Next t
  For t = -ymin To -ymax Step -10
   Picture1.Line (xmin, t)-(xmin + (xmax - xmin) / 40, t)
  Next t

  tetap = Atn(n2 / n1)
  tetap = tetap * 180 / pi
  If n1 < n2 Then GoTo 10 Else GoTo 20
10 'Soft incidence
  For a = xmin + 0.00001 To tetap Step (xmax - xmin) / steps
    fitm = -pi
    Picture1.PSet (a, 180 * fitm / pi), RGB(0, 0, 255)
    fite = 0
    Picture1.PSet (a, 180 * fite / pi), RGB(255, 0, 0)
  Next
  For a = tetap To xmax - 0.000001 Step (xmax - xmin) / steps
    fitm = 0
    Picture1.PSet (a, 180 * fitm / pi), RGB(0, 0, 255)
    fite = 0
```

```
      Picture1.PSet (a, 180 * fite / pi), RGB(255, 0, 0)
    Next
    GoTo 100
 20 'Hard incidence
   tetac = Atn(n2 / (n1 ^ 2 - n2 ^ 2) ^ 0.5)
   tetac = tetac * 180 / pi
   For a = xmin + 0.00001 To tetap Step (xmax - xmin) / steps
     fitm = -pi
     Picture1.PSet (a, 180 * fitm / pi), RGB(0, 0, 255)
     fite = 0
     Picture1.PSet (a, 180 * fite / pi), RGB(255, 0, 0)
   Next
   For a = tetap To tetac - 0.000001 Step (xmax - xmin) / steps
     fitm = 0
     Picture1.PSet (a, 180 * fitm / pi), RGB(0, 0, 255)
     fite = 0
     Picture1.PSet (a, 180 * fite / pi), RGB(255, 0, 0)
   Next
   For a = tetac + 0.000001 To xmax - 0.0001 Step (xmax - xmin) /
   steps tetai = a * pi / 180
     cosenoti = (1 - Sin(tetai) ^ 2) ^ 0.5
     senot = n1 * Sin(tetai) / n2
     B = -(senot ^ 2 - 1) ^ 0.5
     tanfitm = n1 * B / (n2 * cosenoti)
     fitm = 2 * Atn(tanfitm)
     Picture1.PSet (a, 180 * fitm / pi), RGB(0, 0, 255)
     tanfite = n2 * B / (n1 * cosenoti)
     fite = 2 * Atn(tanfite)
     Picture1.PSet (a, 180 * fite / pi), RGB(255, 0, 0)
   Next
100 END
```

Appendix 3

MARCATILI'S METHOD FOR SOLVING GUIDED MODES IN RECTANGULAR CHANNEL WAVEGUIDES

The following sentences are the code for solving quasi-TE guided modes in rectangular waveguides by the Marcatili's method. The integers *p%* and *q%* denote the mode label, and *beta* is the propagation constant of the mode.

```
First Subroutine
      Picture1.Cls
      pi = 4 * Atn(1)
      n1 = Val(Text3.Text)         'Core refractive index (region I)
      n2 = Val(Text4.Text)         'Refractive index of region II
      n3 = Val(Text5.Text)         'Refractive index of region III
      n4 = Val(Text6.Text)         'Refractive index of region IV
      n5 = Val(Text7.Text)         'Refractive index of region V
      lan = Val(Text8.Text)        'Wavelength
      k = 2 * pi / lan
      a = Val(Text9.Text)          'Channel dimension in x-direction
      b = Val(Text10.Text)         'Channel dimension in y-direction
      deltaa = 0.05
      deltab = 0.05


      Picture1.Scale (0, -10)-(n1, 10)
      Picture1.Line (0, 0)-(n1, 0), 1
      Picture1.Line (0, -10)-(n1, 10), 1, B
      acum = 100
      mx% = 0: my% = 0
      For neff = 0.0000001 To n1 Step 0.00005
       kxx = neff * k
       g2p = ((n1 ^ 2 - n2 ^ 2) * k ^ 2 - kxx ^ 2)
       If g2p < 0 Then GoTo 150
       g2 = g2p ^ 0.5
```

```
        g3p = ((n1 ^ 2 - n3 ^ 2) * k ^ 2 - kxx ^ 2)
        If g3p < 0 Then GoTo 150
        g3 = g3p ^ 0.5
        x = Tan(kxx * a)
        y = (n1 ^ 2 * kxx * (n3 ^ 2 * g2 + n2 ^ 2 * g3)) /
(n3 ^ 2 * n2 ^ 2 * kxx ^ 2 - n1 ^ 4 * g2 * g3)
        psi = (1 / kxx) * Atn(-kxx * n3 ^ 2 / (n1 ^ 2 * g3))
        dif = x - y
        If acum <= 0 And dif > 0 Then GoSub 100
        acum = dif
        Picture1.PSet (neff, dif), RGB(0, 0, 255)
        Next neff
        GoTo 150

100       kx(mx%) = kxx
        fx(mx%) = psi
        gx(mx%) = g2
        hx(mx%) = g3

        mx% = mx% + 1
        Return

150
        acum = 100
        For neff = 0.0000001 To n1 Step 0.00005
         kyy = neff * k
         g4p = ((n1 ^ 2 - n4 ^ 2) * k ^ 2 - kyy ^ 2)
         If g4p < 0 Then GoTo 300
         g4 = g4p ^ 0.5
         g5p = ((n1 ^ 2 - n5 ^ 2) * k ^ 2 - kyy ^ 2)
         If g5p < 0 Then GoTo 300
         g5 = g5p ^ 0.5
         x = Tan(kyy * b)
         y = (kyy * (g4 + g5)) / (kyy ^ 2 - g4 * g5)
         eta = (1 / kyy) * Atn(-g5 / kyy)
         dif = x - y
         If acum <= 0 And dif > 0 Then GoSub 200
         acum = dif
         Picture1.PSet (neff, dif), RGB(255, 0, 0)
        Next neff

        GoTo 300

200       ky(my%) = kyy
        fy(my%) = eta
        gy(my%) = g4
        hy(my%) = g5

        my% = my% + 1
        Return
300      'End of first calculation
        Text11.Text = mx% - 1
```

```
    Text12.Text = my% - 1
    Command5.Enabled = False
  End Sub
```

Second Subroutine

```
 '*********DRAW THE MODE *******
   Command5.Enabled = False

    p% = Val(Text1.Text)
    q% = Val(Text2.Text)
    psi = fx(p%)
    kxx = kx(p%)
    eta = fy(q%)
    kyy = ky(q%)
    beta = (n1 ^ 2 * k ^ 2 - (kxx ^ 2 + kyy ^ 2)) ^ 0.5
    g2 = gx(p%): g3 = hx(p%)
    g4 = gy(q%): g5 = hy(q%)
    Picture2.Cls

    neff = beta * lan / (2 * pi)
    Text14.Text = neff
    lonx = Val(Text13.Text) / 2
    lony = Val(Text15.Text) / 2

    Picture2.Scale (b / 2 - lony, -a / 2 + lonx)-(b / 2 + lony, a / 2 - lonx)

    maximum = -1

    '********* Region 1 (core) ********
    prefactor = (n1 ^ 2 * k ^ 2 - kxx ^ 2) / (kxx * beta)
    For x = -a To 0 Step deltaa
     For y = 0 To b Step deltab
      ex = prefactor * Sin(kxx * (x + psi)) * Cos(kyy * (y + eta))
      If Abs(ex) > maximum Then maximum = Abs(ex)
     Next y
    Next x

    '******* Region 2 ******
    For x = -2 * a To -a Step deltaa
     For y = 0 To b Step deltab
      ex = (g2 ^ 2 + n2 ^ 2 * k ^ 2) / (g2 * beta) * Cos(kxx * (psi
- a)) * Cos(kyy * (y + eta)) * Exp(g2 * (x + a))
      If Abs(ex) > maximum Then maximum = Abs(ex)
     Next y
    Next x

    '******* Region 3 ******
    For x = 0 To a Step deltaa
     For y = 0 To b Step deltab
      ex = -(g3 ^ 2 + n3 ^ 2 * k ^ 2) / (g3 * beta) * Cos(kxx * psi)
* Cos(kyy * (y + eta)) * Exp(-g3 * x)
      If Abs(ex) > maximum Then maximum = Abs(ex)
```

```
    Next y
   Next x

   '******** Region 4 *******
   For x = -a To 0 Step deltaa
    For y = b To 2 * b Step deltab
     ex = (n1 ^ 2/n4 ^ 2) * (n4 ^ 2 * k ^ 2-kxx ^ 2)/(kxx * beta) *
Cos(kyy * (b + eta)) * Sin(kxx * (x + psi)) * Exp(-g4 * (y - b))
     If Abs(ex) > maximum Then maximum = Abs(ex)
    Next y
   Next x

   '****** Region 5 *****
   For x = -a To 0 Step deltaa
    For y = -b To 0 Step deltab
     ex = (n1 ^ 2/n5 ^ 2) * (n5 ^ 2 * k ^ 2 - kxx ^ 2) /
(kxx * beta) * Cos(kyy * eta) * Sin(kxx * (x + psi)) * Exp(g5 * y)
     If Abs(ex) > maximum Then maximum = Abs(ex)
    Next y
   Next x

   factor = 255 / maximum ^ 2
   '******* Region 1 ******
   For x = -a To 0 Step deltaa
    For y = 0 To b Step deltab
     ex = prefactor * Sin(kxx * (x + psi)) * Cos(kyy * (y + eta))
     col = Int(factor * ex ^ 2)
     Picture2.Line (y, x)-(y + deltab, x + deltaa), RGB(256 - col,
256 - col, 256 - col), BF
    Next y
   Next x

   '******* Region 2 ******
   For x = -2 * a To -a Step deltaa
    For y = 0 To b Step deltab
     ex = (g2 ^ 2 + n2 ^ 2 * k ^ 2) / (g2 * beta) * Cos(kxx * (psi
- a)) * Cos(kyy * (y + eta)) * Exp(g2 * (x + a))
     col = Int(factor * ex ^ 2)
     Picture2.Line (y, x)-(y + deltab, x + deltaa), RGB(256 - col,
256 - col, 256 - col), BF
    Next y
   Next x

   '******* Region 3 ******
   For x = 0 To a Step deltaa
    For y = 0 To b Step deltab
     ex = -(g3 ^ 2 + n3 ^ 2 * k ^ 2) / (g3 * beta) * Cos(kxx * psi)
* Cos(kyy * (y + eta)) * Exp(-g3 * x)
     col = Int(factor * ex ^ 2)
     Picture2.Line (y, x)-(y + deltab, x + deltaa), RGB(256 - col,
256 - col, 256 - col), BF
```

```
    Next y
   Next x

   '******** Region 4 *******
   For x = -a To 0 Step deltaa
    For y = b To 2 * b Step deltab
     ex = (n1 ^ 2 / n4 ^ 2) * (n4 ^ 2 * k ^ 2 - kxx ^ 2) / (kxx * beta) *
Cos(kyy * (b + eta)) * Sin(kxx * (x + psi)) * Exp(-g4 * (y - b))
     col = Int(factor * ex ^ 2)
     Picture2.Line (y, x)-(y + deltab, x + deltaa), RGB(256 - col,
256 - col, 256 - col), BF
    Next y
   Next x

   '****** Region 5 *****
   For x = -a To 0 Step deltaa
    For y = -b To 0 Step deltab
     ex = (n1 ^ 2 / n5 ^ 2) * (n5 ^ 2 * k ^ 2 - kxx ^ 2)/(kxx *
beta) * Cos(kyy * eta) * Sin(kxx * (x + psi)) * Exp(g5 * y)
     col = Int(factor * ex ^ 2)
     Picture2.Line (y, x)-(y + deltab, x + deltaa), RGB(256 - col,
256 - col, 256 - col), BF
    Next y
   Next x

     Picture2.Line (b, 0)-(0, -a), RGB(255, 0, 0), B

  If Check1.Value = False Then GoTo 400
  'Draw the GRID
  Picture2.Line (b, 0)-(0, -a), RGB(255, 0, 0), B
   For x = -20 To 20
    Picture2.Line (x, -20)-(x, 20), 15
   Next
   For y = -20 To 20
    Picture2.Line (-20, y)-(20, y), 15
  Next

400
End Sub
```

Appendix 4

DEMONSTRATION OF FORMULA (4.3)

Let us assume that two electromagnetic fields ($\mathbf{E}_1$,$\mathbf{H}_1$) and ($\mathbf{E}_2$,$\mathbf{H}_2$) are monochromatic waves propagating along a structure characterised by its optical constants μ_0 and ε. These two fields satisfy Maxwell's equations:

$$\nabla \times \mathbf{E}_1 = -i\omega\mu_0\mathbf{H}_1 \tag{A.1}$$

$$\nabla \times \mathbf{H}_1 = i\omega\varepsilon\mathbf{E}_1 \tag{A.2}$$

$$\nabla \times \mathbf{E}_2 = -i\omega\mu_0\mathbf{H}_2 \tag{A.3}$$

$$\nabla \times \mathbf{H}_2 = i\omega\varepsilon\mathbf{E}_2 \tag{A.4}$$

By combining the above equations, we form the following expression:

$$\mathbf{E}_1^*(\text{A.4})^* - \mathbf{H}_2(\text{A.1}) - \mathbf{H}_1^*(\text{A.3}) + \mathbf{E}_2(\text{A.2})^* \tag{A.5}$$

After straightforward calculation we obtain:

$$\begin{aligned} &\mathbf{E}_1^*\nabla \times \mathbf{H}_2 - \mathbf{H}_2(\nabla \times \mathbf{E}_1)^* - \mathbf{H}_1^*\nabla \times \mathbf{E}_2 + \mathbf{E}_2(\nabla \times \mathbf{H}_1)^* \\ &= \mathbf{E}_1^*(i\omega\varepsilon\mathbf{E}_2) - \mathbf{H}_2(-i\omega\mu_0\mathbf{H}_1)^* - \mathbf{H}_1^*(-i\omega\mu_0\mathbf{H}_2) + \mathbf{E}_2(i\omega\varepsilon\mathbf{E}_1)^* = 0 \end{aligned} \tag{A.6}$$

Now, taking into account the vectorial identity:

$$\nabla(\mathbf{A} \times \mathbf{B}) \equiv \mathbf{B}(\nabla \times \mathbf{A}) - \mathbf{A}(\nabla \times \mathbf{B}) \tag{A.7}$$

the equation (A.6) takes the final form:

$$\nabla(\mathbf{E}_1 \times \mathbf{H}_2^* + \mathbf{E}_2^* \times \mathbf{H}_1) = 0 \tag{A.8}$$

Appendix 5

DERIVATION OF FORMULA (4.4)

Let us assume that $\mathbf{A}(x, y, z)$ is a vectorial function that fulfils the condition:

$$\nabla \mathbf{A}(x, y, z) = 0 \tag{A.9}$$

We perform an integration of the function $\nabla \mathbf{A}$ over the volume inside a cylinder delimited by two circular surfaces perpendicular to the z-axis, as shown in Figure A.1, by evaluating the expression:

$$\iiint_V \nabla \mathbf{A}\, dV \tag{A.10}$$

Making use of the Gauss' theorem, this integral can be converted into a surface integral over the close surface that surrounds the volume V:

$$\iiint_V (\nabla \mathbf{A})\, dV = \oint\!\!\oint_S \mathbf{A}\, d\mathbf{S} = 0 \tag{A.11}$$

The close surface integral in (A.11) can be separated into three parts, corresponding to the two perpendicular surfaces to the z-axis (S_1 and S_2), and the lateral surface of the cylinder (S_3):

$$\oint\!\!\oint_S \mathbf{A}\, d\mathbf{S} = \int_{S1} \mathbf{A}\, d\mathbf{S} + \int_{S2} \mathbf{A}\, d\mathbf{S} + \int_{S3} \mathbf{A}\, d\mathbf{S} = 0 \tag{A.12}$$

If the radius of the cylinder base tends to infinity ($S_1, S_2 \to \infty$), and the cylinder height is much smaller that the cylinder radius, the integral corresponding to the surface S_3 can be neglected, and therefore expression (A.12) simplifies to:

$$\oint\!\!\oint_S \mathbf{A}\, dS = -\int_{S1} A_z(x, y, z)\, dx\, dy + \int_{S2} A_z(x, y, z + \Delta z)\, dx\, dy = 0 \tag{A.13}$$

where A_z indicates the longitudinal component of the vector $\mathbf{A}$.

If the distance Δz tends to zero ($\Delta z \to dz$), equation (A.13) can be converted in:

$$\int_S \frac{\partial}{\partial z} A_z(x, y, z)\, dx\, dy = 0 \tag{A.14}$$

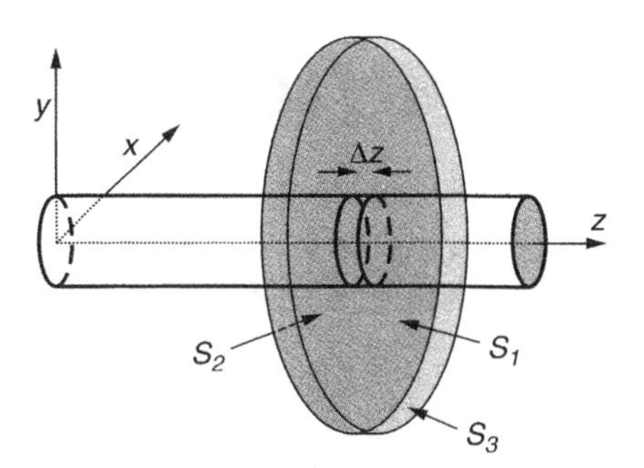

Figure A.1

Now, if the vectorial function **A** stands for the vectorial function:

$$\mathbf{A} \equiv (\mathbf{E}_1 \times \mathbf{H}_2^* + \mathbf{E}_2^* \times \mathbf{H}_1) \tag{A.15}$$

then the following equation is obtained:

$$\int_S \frac{\partial}{\partial z}[\mathbf{E}_1 \times \mathbf{H}_2^* + \mathbf{E}_2^* \times \mathbf{H}_1]_z \, dx \, dy = 0 \tag{A.16}$$

Appendix 6

FAST FOURIER ALGORITHM

Replaces **Dat()** by its discrete Fourier transform, if **isign%** is input as 1; or replaces **Dat()** by **nn%** times its inverse discrete Fourier transform, if **isign%** is input as −1. **Dat()** is a complex array of length **nn%** or, equivalently, a real array of length 2***nn%**. **nn%** must be an integer power of 2. This algorithm can be used to implement the FFT-BPM program, based on (5.16)–(5.19) formulae.

```
Sub FFT()
'**** SUBROUTINE FFT *****
      Dim nn%, j%, i%, m%, mmax%, istep%
      Dim tempr, tempi, wpr, wpi, wr, wi, wtemp
      nn% = 2 * n%
      j% = 1
      For i% = 1 To nn% Step 2
            If j% > i% Then
            tempr = dat(j%)            'Real part of  Dat()
            tempi = dat(j% + 1)        'Imaginary part of  Dat()
            dat(j%) = dat(i%)
            dat(j% + 1) = dat(i% + 1)
            dat(i%) = tempr
            dat(i% + 1) = tempi
            End If
            m% = nn% / 2
1000        if m% >= 2 And j% > m% Then
            j% = j% - m%
            m% = m% / 2
            GoTo 1000
            End If
            j% = j% + m%
       Next i%
       mmax% = 2
2000   If nn% > mmax% Then
             istep% = 2 * mmax%
             teta = 2 * pi / (isign% * mmax%)
             wpr = -2 * (Sin(teta / 2)) ^ 2
             wpi = Sin(teta)
             wr = 1
             wi = 0
             For m% = 1 To mmax% Step 2
```

```
                    For i% = m% To nn% Step istep%
                          j% = i% + mmax%
                          tempr = wr * dat(j%) - wi * dat(j% + 1)
                          tempi = wr * dat(j% + 1) + wi * dat(j%)
                          dat(j%) = dat(i%) - tempr
                          dat(j% + 1) = dat(i% + 1) - tempi
                          dat(i%) = dat(i%) + tempr
                          dat(i% + 1) = dat(i% + 1) + tempi
                          Next i%
                          wtemp = wr
                          wr = wr * wpr - wi * wpi + wr
                          wi = wi * wpr + wtemp * wpi + wi
                     Next m%
                     mmax% = istep%
                    GoTo 2000
          End If
End Sub
```

(Adapted from: W.H. Press, S.A. Teukolsky, W.T. Vetterling and B.P. Flannery, *Numerical Recipes in Fortran 77: The Art of Scientific Computing*, Chapter 12. Cambridge University Press, New York, 1996.)

Appendix 7

IMPLEMENTATION OF THE CRANK-NICOLSON PROPAGATION SCHEME

Implementation of the Crank-Nicolson propagation scheme using the Thomas Method for solving the tridiagonal system described by equations (5.26) and (5.27).

The complex optical field u and other variables (a, b, c and r) are split in their real and imaginary parts as:

$$u = u1 + iu2$$

$$a = a1 + ia2$$

$$b = b1 + ib2$$

$$c = c1 + ic2$$

$$r = r1 + ir2$$

i being the imaginary unity.

The following code solves a propagation step Δz for the optical field u, using Dirichlet boundary conditions if TBC% $= 0$ or transparent boundary conditions if TBC% $= 1$. The integer $n\%$ denotes the number of discretisation points for the optical field.

```
Sub THOMAS()
      Dim q%, s%, j%
      Dim aa, bb, modulo, fi, aa2, bb2, a, bp1, bp2, be1, be2,
t1, t2
      'Transparent Boundary Condition for the right frontier
      q% = 1
      aa = (u1(n% - q%) * u1(n% - 1 - q%) + u2(n% - q%) *
u2(n% - 1 - q%)) / (u1(n% - 1 - q%) ^ 2 + u2(n% - 1 - q%) ^ 2)
      bb = (u1(n% - q%) * u2(n% - 1 - q%) - u2(n% - q%) *
u1(n% - 1 - q%)) / (u1(n% - 1 - q%) ^ 2 + u2(n% - 1 - q%) ^ 2)
      modulo = (aa ^ 2 + bb ^ 2) ^ 0.5
      fi = -Atn(bb / aa)
      aa = modulo * Cos(fi)
      bb = modulo * Sin(fi)
```

```
      If bb > 0 Then aa = modulo: bb = 0
      'Transparent Boundary Condition for the left frontier
      q% = 3: s% = q% - 1
      aa2 = (u1(q%) * u1(s%) + u2(q%) * u2(s%)) /
(u1(s%) ^ 2 + u2(s%) ^ 2)
      bb2 = (u1(q%) * u2(s%) - u2(q%) * u1(s%)) /
(u1(s%) ^ 2 + u2(s%) ^ 2)
      modulo = (aa2 ^ 2 + bb2 ^ 2) ^ 0.5
      fi = -Atn(bb2 / aa2)
      aa2 = modulo * Cos(fi)
      bb2 = modulo * Sin(fi)
      If bb2 < 0 Then aa2 = modulo: bb2 = 0
      'Thomas Algorithm
      a = dz / (2 * dx ^ 2)
      a1(1) = 0: a2(1) = 0
      b1(1) = aa2: b2(1) = bb2
      c1(1) = -1: c2(1) = 0
      a1(n%) = aa: a2(n%) = bb
      b1(n%) = -1: b2(n%) = 0
      c1(n%) = 0: c2(n%) = 0
      r1(1) = 0: r2(1) = 0
      r1(n%) = 0: r2(n%) = 0
      For j% = 1 + TBC% To n% - TBC%
             a1(j%) = -a * alfa: a2(j%) = 0
             b1(j%) = 2 * a * alfa - (dz / 2) * alfa *
(ri(j%) ^ 2 - ri0 ^ 2) * k0 ^ 2
             b2(j%) = k
             c1(j%) = -a * alfa: c2(j%) = 0
             bp1 = -2 * a * (1 - alfa) + (dz / 2) * (1 - alfa) *
(ri(j%) ^ 2 - ri0 ^ 2) * k0 ^ 2
             bp2 = k
             r1(j%) = a * (1 - alfa) * u1(j% - 1) + bp1 * u1(j%) -
bp2 * u2(j%) + a * (1 - alfa) * u1(j% + 1)
             r2(j%) = a * (1 - alfa) * u2(j% - 1) + bp2 * u1(j%) +
bp1 * u2(j%) + a * (1 - alfa) * u2(j% + 1)
      Next j%
      be1 = b1(1): be2 = b2(1)
      u1(1) = (r1(1) * be1 + r2(1) * be2) / (be1 ^ 2 + be2 ^ 2)
      u2(1) = (r2(1) * be1 - r1(1) * be2) / (be1 ^ 2 + be2 ^ 2)
      For j% = 2 To n% Step 1
             g1(j%) = (c1(j% - 1) * be1 + c2(j% - 1) * be2) /
(be1 ^ 2 + be2 ^ 2)
             g2(j%) = (c2(j% - 1) * be1 - c1(j% - 1) * be2) /
(be1 ^ 2 + be2 ^ 2)
             be1 = b1(j%) - a1(j%) * g1(j%) + a2(j%) * g2(j%)
             be2 = b2(j%) - a1(j%) * g2(j%) - a2(j%) * g1(j%)
             If (be1 ^ 2 + be2 ^ 2) < 1E-20 Then End
             t1 = r1(j%) - a1(j%) * u1(j% - 1) + a2(j%) *
u2(j% - 1)
             t2 = r2(j%) - a1(j%) * u2(j% - 1) - a2(j%) *
u1(j% - 1)
             u1(j%) = (t1 * be1 + t2 * be2) /
```

```
(be1 ^ 2 + be2 ^ 2)
            u2(j%) = (t2 * be1 - t1 * be2) /
(be1 ^ 2 + be2 ^ 2)
      Next j%
      For j% = n% - 1 To 1 Step -1
            u1(j%) = u1(j%) - g1(j% + 1) * u1(j% + 1) +
g2(j% + 1) * u2(j% + 1)
            u2(j%) = u2(j%) - g1(j% + 1) * u2(j% + 1) -
g2(j% + 1) * u1(j% + 1)
      Next
End Sub
```

Appendix 8

LIST OF ABBREVIATIONS

AO Acousto-optic
AOTF Acousto-optic tuneable filter
AWG Arrayed waveguide grating
BPM Beam propagation method
CATV Cable television
CMT Coupled mode theory
CVD Chemical vapour deposition
DBR Distributed Bragg reflector
DFB Distributed feedback
DWDM Dense WDM
ECR Electron cyclotron resonance
EIM Effective index method
EM Electromagnetic
EO Electro-optic
FD Finite differences
FFT Fast Fourier transform
FHD Flame hydrolysis deposition
IWKB Inverse WKB
LED Light emitting diode
LPE Liquid phase epitaxy
LSC Luminescent solar concentrator
MBE Molecular beam epitaxy
MMI Multi-mode interference
MOCVD Metal-organic chemical vapour deposition
MZI Mach-Zehnder interferometer
OD Optical density
OPO Optical parametric oscillator
OTDM Optical TDM
PBS Polarisation beam splitter
PHASAR Phase array
PLC Planar lightwave circuit
RF Radio frequency
SAW Surface acoustic wave
TBC Transparent boundary condition
TDM Time division multiplexing
TE Transverse electric
TEM Transversal electromagnetic
TM Transverse magnetic
TO Thermo-optic
WDM Wavelength division multiplexing
WGR Waveguide grating router
WKB Wentzel-Kramers-Brillouin (approximation)

Appendix 9

SOME USEFUL PHYSICAL CONSTANTS

Quantity	Symbol	Value
Speed of light in vacuum	c	3.00×10^{8} m/s
Dielectric permittivity of the vacuum	ε_0	8.85×10^{-12} F/m
Magnetic permeability of the vacuum	μ_0	$4\pi \times 10^{-7}$ H/m
Planck's constant	h	6.63×10^{-34} Js
Boltzmann's constant	K	1.38×10^{-23} J/K
Stefan-Boltzmann's constant	σ	5.67×10^{-9} W/m^{2}K^4
Avogadro's constant	N_A	6.02×10^{23} mol^{-1}
Elementary charge	e	1.60×10^{-19} C
Electron rest mass	m_e	9.11×10^{-31} Kg
Proton rest mass	m_p	1.67×10^{-27} Kg
Gases constant	R	8.31 J/K mol

INDEX

Printed and bound by CPI Group (UK) Ltd, Croydon, CR0 4YY
16/04/2025
14658473-0004